Chapter 1: Introduction to Dopamine and Its Role in the Brain

Overview of Dopamine as a Neurotransmitter

Dopamine, often described as the "feel-good" neurotransmitter, plays an essential role in the brain's reward and pleasure centers. Beyond its reputation in popular culture for making us feel "happy," dopamine is involved in an array of critical physiological functions, including movement, attention, mood regulation, and the reinforcement of behaviors. As a catecholamine neurotransmitter, dopamine is synthesized within neurons and then released to interact with receptors on neighboring cells, initiating complex biochemical responses that influence thought, emotion, and action.

The brain's dopaminergic system is vital for our day-to-day functioning and overall quality of life. Without dopamine, our ability to experience pleasure, pursue goals, and even move our bodies would be severely compromised. Dopamine's centrality in these core functions explains why imbalances or disruptions in dopamine levels can lead to both mental and physical challenges, spanning disorders such as depression, addiction, and neurodegenerative diseases like Parkinson's. The role of dopamine in these conditions has driven extensive research into the ways we can understand, manipulate, and potentially enhance dopamine's effects in the brain.

Importance in Mood, Motivation, and Reward Pathways

Dopamine is critical in regulating the brain's reward system, which affects our mood and motivation. This reward pathway is an interconnected set of brain regions that drive pleasure, reinforce learning, and ultimately shape behavior. When we engage in activities that are essential for survival—such as eating or socializing—dopamine is released, creating a sense of reward that encourages us to repeat those actions.

The motivation-reward cycle is tightly regulated by dopamine's signaling across several brain regions, including the ventral tegmental area (VTA), nucleus accumbens, and prefrontal cortex. Dopamine neurons in the VTA are activated in response to rewarding stimuli or even the anticipation of a reward. From here, signals are sent to the nucleus accumbens, a central hub of the reward circuit that translates dopamine surges into feelings of motivation and pleasure.

As dopamine binds to receptors in these regions, it creates the sense of enjoyment and satisfaction that reinforces behavior. This reward signaling doesn't only relate to physical rewards like food or comfort; it extends to abstract rewards, such as achieving goals or receiving praise. Understanding this mechanism is central to many therapeutic interventions that aim to correct dysfunctions in dopamine signaling associated with various mental health conditions.

Dopamine Synthesis and Metabolism
Biochemical Pathways of Dopamine Synthesis

Dopamine synthesis begins with the amino acid tyrosine, which we acquire through dietary proteins. Tyrosine is converted to L-DOPA (levodopa) by the enzyme tyrosine hydroxylase, which is the rate-limiting step in dopamine production. L-DOPA is then converted to dopamine through the action of aromatic L-amino acid decarboxylase. This two-step process occurs in dopaminergic neurons, primarily in brain regions like the substantia nigra and VTA, which are densely packed with dopamine-producing cells.

Metabolic Breakdown and the Role of Monoamine Oxidase (MAO)

Once released into the synaptic cleft, dopamine doesn't stay active indefinitely. Dopamine molecules that do not bind to receptors are quickly broken down by enzymes, primarily monoamine oxidase (MAO) and catechol-O-methyltransferase (COMT). MAO enzymes, specifically the MAO-B subtype, play a crucial role in degrading dopamine into inactive compounds. By regulating the breakdown of dopamine, MAO-B helps maintain appropriate levels within the brain, preventing an overload or depletion that could disrupt normal cognitive and motor functions.

Understanding MAO and MAO–B Enzymes
Distinguishing Between MAO–A and MAO–B

MAO enzymes exist in two forms, MAO-A and MAO-B, each with distinct but overlapping roles in breaking down neurotransmitters. MAO-A primarily metabolizes serotonin, norepinephrine, and other monoamines, whereas MAO-B primarily targets dopamine. This specificity makes MAO-B a key player in dopaminergic regulation, particularly in areas of the brain associated with movement and reward.

Functions and Impact of MAO–B on Dopamine Regulation

MAO-B's selective activity on dopamine makes it a focal point for researchers interested in modulating dopamine levels, especially for treating neurodegenerative conditions. MAO-B's activity increases with age, contributing to the gradual decline in dopamine that is often associated with aging. This is why inhibiting MAO-B to preserve dopamine levels can have a neuroprotective effect, particularly in conditions like Parkinson's disease.

The Role of MAO–B in Neurodegeneration

MAO–B's Role in Conditions Like Parkinson's and Alzheimer's

In neurodegenerative diseases like Parkinson's, dopamine-producing neurons in the substantia nigra progressively deteriorate, leading to a decline in dopamine levels. Increased MAO-B activity exacerbates this depletion, accelerating neurodegeneration through oxidative stress. By generating reactive oxygen species during dopamine breakdown, MAO-B contributes to cellular damage, particularly in vulnerable dopaminergic neurons.

In Alzheimer's disease, MAO-B has also been implicated in increasing oxidative stress, which contributes to the pathology. Research indicates that inhibiting MAO-B may slow these neurodegenerative processes by preserving dopamine levels and reducing oxidative damage.

Mechanisms of Oxidative Stress and Dopamine Degradation

Oxidative stress occurs when there's an imbalance between the production of free radicals and the body's ability to neutralize them with antioxidants. Dopamine metabolism by MAO-B produces hydrogen peroxide, a reactive oxygen species that can damage cellular components if not adequately neutralized. Over time, this accumulation of oxidative stress leads to cellular dysfunction and death, particularly in dopamine-rich regions of the brain.

Introduction to MAO-B Inhibitors
Basics of MAO-B Inhibition and Its Clinical Relevance

MAO-B inhibitors are a class of drugs designed to prevent the breakdown of dopamine by blocking the activity of MAO-B enzymes. By inhibiting MAO-B, these drugs preserve dopamine levels in the brain, allowing for improved dopaminergic signaling. Clinically, MAO-B inhibitors are used to treat symptoms of Parkinson's disease, offering both symptomatic relief and, potentially, neuroprotection.

Overview of the Therapeutic Benefits

The primary therapeutic benefit of MAO-B inhibitors is to enhance dopamine availability, alleviating motor symptoms in Parkinson's and potentially stabilizing mood in depressive disorders. By slowing the degradation of dopamine, MAO-B inhibitors help maintain neurotransmitter balance, contributing to better mental clarity, motor function, and emotional regulation. Their ability to selectively inhibit MAO-B without affecting MAO-A minimizes the risk of dietary interactions and severe side effects, making them suitable for long-term management of dopamine-related conditions.

This chapter has introduced dopamine's essential role in the brain, its synthesis and regulation, and the significance of MAO-B in maintaining dopamine balance. As we progress through the book, we will explore the pharmacology of specific MAO-B inhibitors, including Selegiline and Rasagiline, examining their potential as therapeutic agents for enhancing dopaminergic function and neuroprotection.

Biochemical Pathways of Dopamine Synthesis

Dopamine is synthesized through a series of enzymatic reactions that transform the amino acid tyrosine into this critical neurotransmitter. The synthesis begins with the dietary intake of protein, which is broken down into amino acids, including phenylalanine and tyrosine. While phenylalanine can be converted into tyrosine, it is tyrosine that directly serves as the precursor for dopamine.

1. **Tyrosine Hydroxylation**: The first step in dopamine synthesis occurs when the enzyme tyrosine hydroxylase catalyzes the conversion of tyrosine to L-DOPA (levodopa). This reaction is the rate-limiting step, meaning it controls the overall speed of dopamine production. Tyrosine hydroxylase requires cofactors such as oxygen and tetrahydrobiopterin (BH4) to function properly.

2. **Decarboxylation of L-DOPA**: The next step involves the decarboxylation of L-DOPA to produce dopamine. This reaction is facilitated by the enzyme aromatic L-amino acid decarboxylase (AAAD), which removes a carboxyl group from L-DOPA. The production of dopamine occurs primarily in dopaminergic neurons, particularly within the substantia nigra and ventral tegmental area (VTA).

This entire synthesis pathway is tightly regulated, as both excess and deficiency of dopamine can lead to significant neurological and psychiatric disorders. Therefore, understanding how dopamine is synthesized helps in grasping its role in various brain functions and its impact on health.

Metabolic Breakdown and the Role of Monoamine Oxidase (MAO)

Once synthesized and released into the synaptic cleft, dopamine does not remain active indefinitely. Its actions are terminated primarily through reuptake into the presynaptic neuron and metabolic breakdown. The latter is crucial for regulating dopamine levels and preventing potential neurotoxicity from excess dopamine.

1. **Reuptake**: Dopamine transporters (DAT) located on the presynaptic membrane facilitate the reuptake of dopamine from the synaptic cleft. This process allows for the recycling of dopamine and helps maintain appropriate neurotransmitter levels in the synapse.

2. **Metabolism by MAO and COMT: If dopamine is not taken back up, it is subjected to enzymatic degradation. The main enzyme responsible for dopamine metabolism is monoamine oxidase (MAO), specifically the MAO-B isoform in the central nervous system. MAO catalyzes the oxidative deamination of dopamine, converting it into the inactive metabolite dihydroxyphenylacetic acid (DOPAC) and hydrogen peroxide.**

Role of MAO-B

: MAO-B selectively metabolizes dopamine, playing a crucial role in maintaining dopamine homeostasis in the brain. Increased activity of MAO-B can lead to reduced dopamine availability, which is particularly relevant in age-related decline and neurodegenerative conditions.

3. **Catechol-O-Methyltransferase (COMT)**: Another enzyme involved in dopamine metabolism is catechol-O-methyltransferase (COMT), which methylates dopamine and its metabolites, further contributing to the breakdown of catecholamines.

The regulation of dopamine levels through both reuptake and metabolism is essential for proper neurological function. Disruptions in these processes can lead to various health issues, including mood disorders, attention deficits, and neurodegenerative diseases.

Understanding MAO and MAO–B Enzymes
Distinguishing Between MAO–A and MAO–B

Monoamine oxidase exists in two primary forms: MAO-A and MAO-B. While both isoforms catalyze the breakdown of neurotransmitters, they differ in substrate specificity and tissue distribution.

- **MAO-A**: This isoform preferentially metabolizes serotonin and norepinephrine. It is primarily found in the gastrointestinal tract and in the brain. Inhibition of MAO-A can lead to increased levels of serotonin, making it a target for certain antidepressant therapies.

- **MAO-B**: In contrast, MAO-B is predominantly responsible for the metabolism of dopamine. It is more prevalent in the brain, particularly in regions involved in movement and reward. MAO-B's unique role in breaking down dopamine highlights its significance in conditions like Parkinson's disease, where dopamine depletion is a hallmark.

Understanding the differences between these two enzymes is crucial for developing targeted therapies that modulate their activity, particularly for neurodegenerative and mood disorders.

Functions and Impact of MAO–B on Dopamine Regulation

MAO-B plays a significant role in regulating dopamine levels in the brain. As mentioned, increased MAO-B activity can lead to reduced dopamine availability, exacerbating symptoms in conditions like Parkinson's disease. This relationship between MAO-B and dopamine is critical for developing therapeutic strategies that aim to restore dopaminergic balance.

The activity of MAO-B also varies with age and in response to certain environmental factors. Elevated MAO-B levels have been observed in aging individuals and in patients with neurodegenerative diseases, correlating with the progression of these conditions. Consequently, inhibiting MAO-B could potentially provide neuroprotective benefits by preserving dopamine levels and preventing oxidative stress associated with dopamine metabolism.

The Role of MAO-B in Neurodegeneration
MAO-B's Role in Conditions Like Parkinson's and Alzheimer's

In Parkinson's disease, the progressive degeneration of dopaminergic neurons in the substantia nigra leads to a critical reduction in dopamine levels. MAO-B activity is elevated in this condition, contributing to the exacerbation of dopamine depletion through increased oxidative stress. This dual role of MAO-B—both in the breakdown of dopamine and in the promotion of neurotoxicity—highlights its potential as a therapeutic target.

Research has shown that inhibiting MAO-B can not only alleviate the symptoms of Parkinson's disease by increasing dopamine availability but also may offer neuroprotective effects. By reducing oxidative stress, MAO-B inhibitors could potentially slow the progression of neurodegeneration.

In Alzheimer's disease, similar mechanisms are observed. Elevated MAO-B levels contribute to neuronal damage and cognitive decline through oxidative stress and inflammatory processes. Targeting MAO-B in Alzheimer's patients could, therefore, be a promising avenue for research and therapy.

Mechanisms of Oxidative Stress and Dopamine Degradation

Oxidative stress is a significant contributor to the pathology of many neurodegenerative diseases. As dopamine is metabolized by MAO-B, reactive oxygen species, including hydrogen peroxide, are generated. These reactive species can cause cellular damage, leading to neuronal death and dysfunction.

The interplay between dopamine degradation and oxidative stress forms a vicious cycle: increased MAO-B activity leads to more oxidative stress, which in turn exacerbates dopamine neuron loss, further increasing MAO-B activity. This understanding underscores the importance of developing MAO-B inhibitors not only to enhance dopamine signaling but also to protect against oxidative damage in dopaminergic neurons.

Introduction to MAO–B Inhibitors
Basics of MAO–B Inhibition and Its Clinical Relevance

MAO-B inhibitors represent a class of medications designed to prevent the breakdown of dopamine by blocking the action of the MAO-B enzyme. By inhibiting this enzyme, these drugs increase the availability of dopamine, which can help manage symptoms in conditions characterized by dopamine deficiency, such as Parkinson's disease and depression.

The clinical relevance of MAO-B inhibitors extends beyond simple symptom management. By preserving dopamine levels and reducing oxidative stress, these inhibitors may offer neuroprotective benefits, addressing not just the symptoms but also the underlying pathophysiology of neurodegenerative diseases.

Overview of the Therapeutic Benefits

The therapeutic benefits of MAO-B inhibitors include improved motor function in Parkinson's patients, enhanced mood regulation in depressive disorders, and potential cognitive benefits. By modulating dopamine levels, MAO-B inhibitors help restore balance in the dopaminergic system, leading to improved quality of life and functional outcomes for individuals with dopamine-related conditions.

Additionally, ongoing research into the neuroprotective properties of these inhibitors suggests they could play a vital role in the future of neurotherapeutics, particularly in delaying the onset or progression of neurodegenerative diseases.

Chapter 3: Understanding MAO and MAO-B Enzymes

Distinguishing Between MAO-A and MAO-B

Monoamine oxidase (MAO) is an important enzyme family responsible for the degradation of monoamines, which include neurotransmitters such as dopamine, serotonin, and norepinephrine. There are two isoforms of this enzyme: MAO-A and MAO-B. Understanding the distinctions between these two enzymes is crucial for comprehending their respective roles in neurotransmitter metabolism and the potential therapeutic targets they represent.

- **MAO-A**: This isoform preferentially metabolizes serotonin and norepinephrine. It is predominantly found in the brain, liver, and gastrointestinal tract. MAO-A plays a significant role in regulating mood and anxiety levels, which has made it a target for antidepressant therapies. Inhibition of MAO-A can lead to increased levels of serotonin and norepinephrine, providing therapeutic benefits in treating mood disorders.

- **MAO-B**: In contrast, MAO-B primarily metabolizes dopamine and is mainly localized in the brain, particularly in regions associated with movement and reward, such as the striatum and substantia nigra. It also exists in the platelets and is involved in the metabolism of phenethylamine, an endogenous monoamine. The significance of MAO-B in dopamine metabolism makes it a critical target for therapies aimed at conditions like Parkinson's disease, where dopaminergic neurons degenerate and dopamine levels are diminished.

Both MAO-A and MAO-B share similar structures and mechanisms, but they differ in their substrate specificity, tissue distribution, and regulation. This distinction is vital for developing selective inhibitors that can enhance the therapeutic efficacy while minimizing side effects.

Functions and Impact of MAO–B on Dopamine Regulation

MAO-B plays a pivotal role in the regulation of dopamine levels in the brain. As dopamine is released into the synaptic cleft, it interacts with dopamine receptors to exert its effects. However, excess dopamine can lead to neurotoxicity and has been implicated in various neurological disorders. Therefore, MAO-B helps maintain balance by degrading excess dopamine.

Regulation of Dopamine Availability

1. **Oxidative Deamination**: MAO-B catalyzes the oxidative deamination of dopamine, converting it into dihydroxyphenylacetic acid (DOPAC) and hydrogen peroxide. This process not only regulates dopamine levels but also plays a role in producing reactive oxygen species, which can contribute to oxidative stress in the brain.

2. **Neuroprotection vs. Neurotoxicity**: While MAO-B serves to prevent dopamine accumulation, its activity also leads to the production of potentially harmful byproducts. Elevated MAO-B activity is associated with increased oxidative stress, which can contribute to neuronal damage, especially in dopamine-rich areas of the brain. This dual role of MAO-B underscores its importance as a target for therapeutic interventions, particularly in neurodegenerative diseases where dopamine depletion occurs.

The Role of MAO-B in Neurodegeneration
MAO-B's Role in Conditions Like Parkinson's and Alzheimer's

MAO-B has been implicated in several neurodegenerative conditions, most notably Parkinson's disease and Alzheimer's disease. The understanding of its role in these conditions provides insights into potential therapeutic strategies.

- **Parkinson's Disease**: In Parkinson's disease, the progressive loss of dopaminergic neurons in the substantia nigra leads to a significant decrease in dopamine levels. Increased MAO-B activity has been observed in the brains of patients with Parkinson's disease, contributing to the oxidative stress and neuroinflammation that exacerbate dopaminergic cell death. Inhibiting MAO-B can help preserve remaining dopamine levels, improve motor symptoms, and potentially offer neuroprotective effects.
- **Alzheimer's Disease**: Similarly, in Alzheimer's disease, elevated MAO-B levels have been linked to increased oxidative stress and cognitive decline. Research suggests that targeting MAO-B may have therapeutic implications in slowing down the progression of Alzheimer's disease by mitigating oxidative damage and preserving neuronal health.

Mechanisms of Oxidative Stress and Dopamine Degradation

The relationship between MAO-B activity and oxidative stress is complex. As MAO-B breaks down dopamine, it generates hydrogen peroxide, a reactive oxygen species. This process can lead to oxidative stress, which damages cellular components, including lipids, proteins, and DNA.

- **Oxidative Stress and Neuroinflammation**: Chronic oxidative stress is a significant contributor to neuroinflammation, further promoting neuronal damage. In conditions like Parkinson's and Alzheimer's, the interplay between oxidative stress and neuroinflammation creates a detrimental cycle that accelerates disease progression.

- **Importance of Antioxidant Defense**: The brain has its own antioxidant defense mechanisms, but when overwhelmed by excessive oxidative stress from increased MAO-B activity, these defenses may fail. This failure can lead to an increase in the vulnerability of dopaminergic neurons, which are already at risk due to their high metabolic activity and susceptibility to oxidative damage.

Introduction to MAO-B Inhibitors
Basics of MAO-B Inhibition and Its Clinical Relevance

Given the critical role of MAO-B in dopamine metabolism and its implications in neurodegenerative diseases, MAO-B inhibitors have emerged as a valuable therapeutic option. These drugs function by selectively inhibiting the activity of the MAO-B enzyme, thereby preserving dopamine levels and mitigating the effects of oxidative stress.

The clinical relevance of MAO-B inhibitors lies in their ability to enhance dopaminergic signaling in conditions characterized by dopamine deficiency, such as Parkinson's disease. By blocking the breakdown of dopamine, these inhibitors provide symptomatic relief and may offer neuroprotective benefits, making them a crucial component of Parkinson's disease management.

Overview of the Therapeutic Benefits

The therapeutic benefits of MAO-B inhibitors extend beyond simply increasing dopamine levels. These drugs may:

1. **Alleviate Motor Symptoms**: By enhancing dopamine availability, MAO-B inhibitors can improve motor function in patients with Parkinson's disease, reducing symptoms such as tremors, rigidity, and bradykinesia.

2. **Provide Neuroprotective Effects**: Some studies suggest that MAO-B inhibitors may have neuroprotective properties, potentially slowing the progression of neurodegenerative diseases by reducing oxidative stress and preventing neuronal death.

3. **Improve Mood and Cognitive Function**: In addition to their effects on motor symptoms, MAO-B inhibitors have been shown to improve mood and cognitive function in some patients, highlighting their potential utility in treating depression and other mood disorders.

4. **Minimize Dietary Restrictions**: Unlike non-selective MAO inhibitors, which can interact with certain foods (especially those containing tyramine), MAO-B inhibitors allow for greater dietary freedom, making them more tolerable for long-term use.

Chapter 4: The Role of MAO-B in Neurodegeneration

MAO-B's Role in Conditions Like Parkinson's and Alzheimer's

Monoamine oxidase B (MAO-B) has garnered considerable attention due to its significant role in the pathology of various neurodegenerative diseases, most notably Parkinson's disease and Alzheimer's disease. Understanding how MAO-B influences these conditions provides crucial insights into potential therapeutic interventions that can mitigate neuronal loss and improve patient outcomes.

Parkinson's Disease: A Closer Look at MAO-B

Parkinson's disease is characterized by the progressive degeneration of dopaminergic neurons in the substantia nigra, leading to a critical reduction in dopamine levels. This depletion manifests in several motor symptoms, including tremors, rigidity, bradykinesia, and postural instability. As the disease progresses, non-motor symptoms such as depression, cognitive decline, and autonomic dysfunction may also arise.

1. **Increased MAO-B Activity**: Research has shown that MAO-B activity is elevated in the brains of individuals with Parkinson's disease. This increased activity exacerbates the loss of dopaminergic neurons through heightened oxidative stress and neuroinflammation. The oxidative deamination of dopamine by MAO-B generates reactive oxygen species, which can lead to cellular damage and neuronal death.

2. **Neuroprotective Strategies**: Inhibiting MAO-B has emerged as a promising therapeutic strategy for managing Parkinson's disease. By blocking MAO-B activity, these inhibitors help preserve dopamine levels, potentially alleviating motor symptoms and providing neuroprotective benefits. Studies suggest that MAO-B inhibitors may slow the progression of the disease by reducing oxidative damage and promoting the survival of remaining dopaminergic neurons.

Alzheimer's Disease and the Role of MAO-B

Alzheimer's disease, another common neurodegenerative disorder, is characterized by progressive cognitive decline, memory loss, and behavioral changes. While the primary pathological features of Alzheimer's include amyloid plaques and neurofibrillary tangles, the role of MAO-B in this condition is increasingly being recognized.

1. **Elevated MAO-B in Alzheimer's**: Similar to Parkinson's disease, MAO-B activity is elevated in the brains of individuals with Alzheimer's disease. This elevation contributes to increased oxidative stress, which is thought to exacerbate neuronal damage and cognitive decline. The relationship between MAO-B activity and the pathophysiology of Alzheimer's highlights the importance of targeting this enzyme for potential therapeutic benefits.

2. **Potential Neuroprotective Effects**: Inhibiting MAO-B in Alzheimer's patients may help reduce oxidative stress and promote neuronal health. Early research suggests that MAO-B inhibitors could improve cognitive function and slow disease progression, although more studies are needed to confirm these effects and establish optimal treatment protocols.

Mechanisms of Oxidative Stress and Dopamine Degradation

The interplay between MAO-B activity, oxidative stress, and dopamine degradation is critical in understanding the neurodegenerative processes involved in Parkinson's and Alzheimer's diseases.

The Role of Reactive Oxygen Species

As MAO-B metabolizes dopamine, it produces reactive oxygen species (ROS), including hydrogen peroxide. Under normal conditions, the brain's antioxidant systems can manage ROS levels; however, excessive production leads to oxidative stress, which damages cellular structures, proteins, and DNA.

1. **Neuronal Vulnerability**: Dopaminergic neurons are particularly vulnerable to oxidative damage due to their high metabolic activity and the presence of dopamine, which can auto-oxidize and generate additional free radicals. This vulnerability contributes to the pathogenesis of both Parkinson's and Alzheimer's diseases.

2. **Cascading Effects**: The oxidative stress generated by MAO-B activity can initiate a cascade of neuroinflammatory responses, further contributing to neuronal damage. Activated microglia and astrocytes release pro-inflammatory cytokines, exacerbating the neurodegenerative process.

3. **Cognitive Decline and Neuroinflammation**: In Alzheimer's disease, the combination of oxidative stress and neuroinflammation is thought to play a significant role in cognitive decline. The presence of amyloid-beta plaques and tau tangles, in conjunction with elevated MAO-B levels, creates an environment that fosters neuronal death and impairs synaptic function.

Introduction to MAO-B Inhibitors

Given the critical role of MAO-B in neurodegeneration, the development of MAO-B inhibitors has become a focal point in therapeutic research. These inhibitors aim to preserve dopamine levels and protect neurons from oxidative stress and inflammation.

Basics of MAO-B Inhibition

MAO-B inhibitors work by selectively binding to the MAO-B enzyme, preventing it from metabolizing dopamine. This inhibition leads to increased availability of dopamine in the synaptic cleft, alleviating symptoms associated with dopamine deficiency.

1. **Clinical Relevance**: The clinical application of MAO-B inhibitors has shown promising results in managing symptoms of Parkinson's disease. By enhancing dopaminergic transmission, these drugs can improve motor function and reduce disability in patients.

2. **Neuroprotective Potential**: Beyond symptomatic relief, MAO-B inhibitors may provide neuroprotective effects by reducing oxidative stress and promoting neuronal survival. This dual action positions them as critical components in the treatment paradigm for neurodegenerative diseases.

Overview of the Therapeutic Benefits

The therapeutic benefits of MAO-B inhibitors are multifaceted and extend beyond merely increasing dopamine availability. These drugs may offer:

1. **Symptomatic Relief in Parkinson's**: MAO-B inhibitors are effective in improving motor symptoms and overall quality of life in Parkinson's disease patients. They are often used as adjunctive therapy alongside dopaminergic medications.

2. **Neuroprotection**: By mitigating oxidative stress and inflammatory responses, MAO-B inhibitors may slow the progression of neurodegenerative diseases, offering long-term benefits for patients.

3. **Cognitive Enhancement**: Some studies suggest that MAO-B inhibitors could also have cognitive benefits, particularly in Alzheimer's disease, by enhancing synaptic function and reducing cognitive decline.

Chapter 5: Pharmacology of Selegiline

Mechanism of Action for Selegiline

Selegiline, also known as L-deprenyl, is a selective inhibitor of monoamine oxidase B (MAO-B). It was initially developed in the 1960s as an antidepressant but gained prominence for its neuroprotective effects and ability to manage symptoms of Parkinson's disease. Understanding its pharmacological mechanisms provides insight into its therapeutic benefits and potential applications.

1. **Selective Inhibition of MAO-B**: Selegiline exhibits a high affinity for MAO-B over MAO-A, which allows it to inhibit the breakdown of dopamine without significantly affecting the metabolism of serotonin and norepinephrine. This selectivity is crucial for its role in enhancing dopaminergic activity while minimizing the risk of side effects associated with non-selective MAO inhibitors.

2. **Enhancement of Dopamine Availability**: By inhibiting MAO-B, Selegiline prevents the enzymatic degradation of dopamine, leading to increased levels of this neurotransmitter in the synaptic cleft. This mechanism is particularly beneficial in Parkinson's disease, where dopaminergic neurons are compromised.

3. **Formation of Neuroprotective Metabolites**: Interestingly, Selegiline is metabolized into several active compounds, including amphetamine and methamphetamine, which may have additional stimulant and neuroprotective properties. These metabolites are thought to enhance dopamine release and contribute to the overall efficacy of Selegiline in managing Parkinson's symptoms.

4. **Antioxidant Effects**: Selegiline may also exert antioxidant effects, reducing oxidative stress in the brain. By decreasing the formation of reactive oxygen species that arise from dopamine metabolism, it may protect vulnerable dopaminergic neurons from damage, further supporting its neuroprotective role.

History and Clinical Applications

Selegiline's journey from initial development to clinical use illustrates the evolution of our understanding of dopamine and its associated pathways.

1. **Early Development**: Selegiline was initially synthesized in the 1960s as a potential antidepressant. However, its selective MAO-B inhibition was discovered, leading researchers to investigate its role in Parkinson's disease.

2. **FDA Approval and Use in Parkinson's Disease**: In 1989, Selegiline was approved for use in the United States as an adjunct therapy for Parkinson's disease. Its ability to enhance the effects of levodopa, the primary treatment for Parkinson's, made it a valuable addition to dopaminergic therapy. By delaying the onset of motor fluctuations and reducing the required dose of levodopa, Selegiline significantly improves the quality of life for many patients.

3. **Additional Applications**: Besides its established role in Parkinson's disease, Selegiline has been investigated for various other applications, including its use as an adjunct treatment for major depressive disorder and for cognitive enhancement in elderly patients. The potential neuroprotective effects have led to research into its use in Alzheimer's disease, although findings remain preliminary.

Clinical Indications and Dosing

Selegiline is primarily indicated for:

- **Parkinson's Disease**: Used alone or in combination with other dopaminergic therapies. It is particularly effective in early-stage Parkinson's disease and for patients experiencing motor fluctuations while on levodopa.
- **Major Depressive Disorder**: In some countries, Selegiline is approved for treating depression, particularly in formulations that are delivered via a transdermal patch, which allows for continuous delivery and minimizes dietary restrictions.

Dosage Recommendations

- **Oral Administration**: The standard oral dosage for Selegiline in Parkinson's disease starts at 5 mg once daily, which can be increased to 10 mg if necessary. The optimal dose may vary based on individual response and the presence of other medications.
- **Transdermal Patch**: The transdermal delivery system (e.g., Emsam) provides a unique method of administration, with dosages typically ranging from 6 mg to 12 mg per day, depending on the specific indications and patient needs.

Common Side Effects

While Selegiline is generally well-tolerated, it may be associated with several side effects, including:

1. **Gastrointestinal Disturbances**: Nausea, vomiting, and diarrhea are common side effects that may occur, particularly when treatment is initiated or the dosage is increased.

2. **Insomnia and Agitation**: Some patients may experience difficulty sleeping or increased restlessness, likely due to the amphetamine metabolites that can stimulate the central nervous system.

3. **Orthostatic Hypotension**: Selegiline may lead to low blood pressure upon standing, causing dizziness or fainting in some individuals.

4. **Dietary Considerations**: Unlike non-selective MAO inhibitors, Selegiline has a lower risk of dietary restrictions, though patients should still avoid tyramine-rich foods in high doses, particularly if using higher doses or the transdermal patch.

Conclusion

Selegiline's unique pharmacological properties as a selective MAO-B inhibitor highlight its importance in managing Parkinson's disease and its potential applications in other neurodegenerative disorders. By enhancing dopamine availability and providing neuroprotective effects, Selegiline plays a critical role in the therapeutic landscape for patients struggling with dopamine-related conditions.

In the following chapter, we will explore the pharmacology of Rasagiline, another MAO-B inhibitor, comparing its mechanisms, efficacy, and clinical applications to those of Selegiline. Understanding these differences will further enhance our mastery of dopamine modulation and its implications for treating neurodegenerative diseases.

Chapter 6: Pharmacology of Rasagiline

Rasagiline's Mechanism of Action and Unique Properties

Rasagiline is a selective and irreversible inhibitor of monoamine oxidase B (MAO-B), specifically designed to enhance dopaminergic function in the brain. First introduced in the late 1990s, it has become an important therapeutic option in the management of Parkinson's disease. Understanding its pharmacological properties and mechanisms of action sheds light on its effectiveness and potential applications.

1. **Selective MAO-B Inhibition**: Similar to Selegiline, Rasagiline selectively inhibits MAO-B, which is crucial for the breakdown of dopamine in the brain. By blocking this enzyme, Rasagiline increases dopamine availability, thereby enhancing dopaminergic signaling, which is particularly beneficial in Parkinson's disease where dopamine levels are diminished.

2. **Irreversible Binding**: Rasagiline binds irreversibly to the MAO-B enzyme, leading to prolonged inhibition. This characteristic sets Rasagiline apart from other reversible MAO-B inhibitors, allowing for sustained dopaminergic effects without the need for frequent dosing. This long-lasting action can contribute to improved patient compliance and therapeutic outcomes.

3. **Neuroprotective Effects**: Beyond its role in enhancing dopamine levels, Rasagiline is believed to exert neuroprotective effects through mechanisms that may involve reducing oxidative stress and inflammation. By inhibiting the degradation of dopamine and decreasing the formation of harmful reactive oxygen species, Rasagiline helps protect dopaminergic neurons from damage.

4. **Additional Metabolites**: Rasagiline is metabolized in the liver to an active metabolite called aminoindan, which also exhibits neuroprotective properties. This dual action—both from Rasagiline and its metabolites—may enhance its therapeutic benefits, particularly in terms of neuronal survival and functional outcomes in Parkinson's disease.

History and Clinical Applications

The clinical journey of Rasagiline reflects its development as a targeted therapy for Parkinson's disease, supported by a growing body of research highlighting its efficacy and safety.

1. **Development Timeline**: Rasagiline was developed in the 1990s and underwent extensive clinical trials to evaluate its safety and efficacy. It was approved for use in several countries, including the United States, where it was sanctioned in 2006 for the treatment of Parkinson's disease.

2. **Indications**: Rasagiline is indicated for the treatment of Parkinson's disease, both as monotherapy in early-stage patients and as an adjunct therapy to levodopa in those experiencing motor fluctuations. Its ability to enhance the effects of levodopa while minimizing the required dosage has made it a valuable option in clinical practice.

3. **Use in Neurodegenerative Diseases**: Research is ongoing to explore Rasagiline's potential applications beyond Parkinson's disease. Investigations into its neuroprotective effects have prompted studies into other neurodegenerative disorders, such as Alzheimer's disease and Huntington's disease, although conclusive clinical results are still pending.

Clinical Indications and Dosing

Rasagiline is primarily indicated for the following:

Parkinson's Disease

: It is effective as a first-line treatment in patients with early Parkinson's disease and as an adjunct therapy for those on levodopa who experience motor fluctuations.

Dosage Recommendations

- **Oral Administration**: The recommended dose of Rasagiline is typically 1 mg once daily. This dosing schedule is straightforward and convenient, contributing to patient adherence.
- **Adjustments for Special Populations**: In cases where patients are taking certain medications that may interact with Rasagiline, or in those with severe hepatic impairment, careful monitoring and potential dose adjustments may be required.

Common Side Effects

While generally well-tolerated, Rasagiline may have some side effects, including:

1. **Headache**: One of the more common side effects, which can occur as the body adjusts to the medication.

2. **Gastrointestinal Symptoms**: Nausea, diarrhea, and abdominal pain may be reported, particularly during the initial phase of treatment.

3. **Insomnia and Dizziness**: Some patients may experience sleep disturbances and dizziness, likely due to the stimulating effects of the drug or its metabolites.

4. **Risk of Serotonin Syndrome**: Given its action on monoamines, caution is advised when combining Rasagiline with other medications that affect serotonin levels, such as SSRIs or certain pain medications.

Comparison with Selegiline

While both Selegiline and Rasagiline are MAO-B inhibitors, several differences and similarities are noteworthy:

1. **Selectivity and Binding**: Both drugs exhibit selectivity for MAO-B, but Rasagiline binds irreversibly, providing a longer duration of action compared to the reversible inhibition seen with Selegiline.

2. **Metabolic Pathways**: Selegiline produces amphetamines as metabolites, which can contribute to its stimulant effects. In contrast, Rasagiline's primary metabolite, aminoindan, is thought to contribute to its neuroprotective effects without the same risk of stimulant-related side effects.

3. **Dosing and Administration**: Rasagiline's once-daily dosing regimen is advantageous for patient compliance, particularly compared to Selegiline, which may require more frequent dosing.

4. **Efficacy in Clinical Trials**: Clinical studies have demonstrated that both drugs are effective in managing Parkinson's disease symptoms; however, individual responses may vary. Rasagiline has also shown evidence of neuroprotective effects in some studies, though further research is needed for definitive conclusions.

Conclusion

Rasagiline represents a significant advancement in the pharmacological management of Parkinson's disease, providing effective dopaminergic support and potential neuroprotective benefits through its selective and irreversible inhibition of MAO-B. Understanding its mechanism of action, clinical applications, and side effect profile allows for better utilization in clinical practice.

In the next chapter, we will compare Selegiline and Rasagiline in detail, exploring their pharmacokinetics, dosing strategies, and efficacy in various patient populations. This comparison will further clarify how these MAO-B inhibitors can be tailored to meet individual patient needs in managing dopamine-related disorders.

Chapter 7: Comparing Selegiline and Rasagiline

In the realm of MAO-B inhibitors, Selegiline and Rasagiline stand out as key therapeutic agents in managing Parkinson's disease and other dopaminergic disorders. Although both drugs share a similar pharmacological goal—enhancing dopamine availability—their mechanisms, pharmacokinetics, dosing regimens, and clinical applications offer important distinctions. This chapter aims to provide a comprehensive comparison of Selegiline and Rasagiline to guide clinical decision-making and optimize patient outcomes.

Pharmacokinetics

Pharmacokinetics refers to how a drug is absorbed, distributed, metabolized, and excreted by the body. Understanding the pharmacokinetic profiles of Selegiline and Rasagiline helps inform clinical use and patient management.

1. **Selegiline:**

   - **Absorption**: Selegiline is rapidly absorbed after oral administration, with peak plasma concentrations typically reached within 1-2 hours. The bioavailability of orally administered Selegiline can vary, especially due to first-pass metabolism in the liver.
   - **Metabolism**: It is extensively metabolized in the liver, primarily by cytochrome P450 enzymes (CYP2B6, CYP2C19). Its metabolic products include both amphetamine and methamphetamine, which may contribute to its stimulant effects.
   - **Half-life**: The elimination half-life of Selegiline ranges from 10 to 12 hours, depending on the formulation and dosage.

2. **Rasagiline:**

- **Absorption**: Rasagiline is also rapidly absorbed, with peak plasma concentrations occurring within 0.5 to 1 hour post-administration. Its bioavailability is higher compared to Selegiline due to reduced first-pass metabolism.
- **Metabolism**: Rasagiline is primarily metabolized to the active metabolite aminoindan, which contributes to its therapeutic effects. It undergoes metabolism mainly via CYP1A2, and unlike Selegiline, it does not produce significant amounts of amphetamines.
- **Half-life**: Rasagiline has a longer half-life of approximately 1.5 to 2 hours, but its irreversible binding to MAO-B results in sustained inhibition of the enzyme, allowing for effective dosing once daily.

Dosing

Dosing regimens for Selegiline and Rasagiline differ, reflecting their pharmacokinetic profiles and clinical indications.

1. **Selegiline:**

   - **Standard Dosing**: The usual starting dose for oral Selegiline in Parkinson's disease is 5 mg once daily, which can be increased to 10 mg based on patient response and tolerability. Selegiline is also available as an orally disintegrating tablet and a transdermal patch, with different dosing considerations for each formulation.
   - **Transdermal Patch**: The patch allows for continuous delivery of Selegiline while minimizing dietary restrictions related to tyramine.

2. **Rasagiline:**

   Standard Dosing

   : Rasagiline is typically administered at a dose of 1 mg once daily, both as monotherapy and in conjunction with levodopa. This straightforward dosing regimen enhances patient compliance, particularly in early-stage Parkinson's disease.

Efficacy

Both Selegiline and Rasagiline have demonstrated efficacy in managing Parkinson's disease, but research indicates some differences in their effectiveness and clinical application.

1. **Selegiline:**

Clinical studies have shown that Selegiline can delay the need for levodopa therapy and improve motor symptoms in early-stage Parkinson's disease. However, its efficacy may diminish over time, particularly as the disease progresses.

2. **Rasagiline:**

Rasagiline has been shown to provide symptomatic relief and may also offer neuroprotective effects in patients with Parkinson's disease. Clinical trials indicate that it may slow disease progression, although results may vary among individuals.

Key Differentiators

1. **Metabolic Profile**: The metabolic products of Selegiline (amphetamine and methamphetamine) can lead to stimulating effects, which may not be suitable for all patients. Rasagiline's lack of significant stimulant metabolites reduces the risk of such side effects, making it preferable for patients who may be sensitive to stimulation.

2. **Duration of Action**: Rasagiline's irreversible binding to MAO-B allows for prolonged enzyme inhibition with a once-daily dosing schedule, providing a practical advantage for long-term management of Parkinson's disease.

3. **Potential for Cognitive Effects**: Some studies suggest that Rasagiline may have additional cognitive benefits due to its neuroprotective properties, an area where ongoing research continues to explore its full potential.

Clinical Use Cases

1. **Early-Stage Parkinson's Disease**: Both Selegiline and Rasagiline are effective as first-line treatments for early-stage Parkinson's disease. The choice between the two may depend on patient-specific factors, including tolerability, response to previous treatments, and concerns about stimulating effects.

2. **Advanced Parkinson's Disease**: For patients with advanced Parkinson's who experience motor fluctuations while on levodopa, Rasagiline may provide an advantageous option due to its sustained effect and reduced side effects compared to Selegiline.

3. **Patient Compliance**: The ease of use associated with Rasagiline's once-daily dosing regimen may improve patient adherence, particularly in elderly populations or those with complex medication regimens.

Conclusion

Selegiline and Rasagiline both play significant roles in the management of Parkinson's disease through their actions as MAO-B inhibitors. While they share a common therapeutic goal of enhancing dopaminergic function, their pharmacokinetic profiles, dosing regimens, and clinical applications present distinct differences that can influence treatment decisions.

In the next chapter, we will delve into the role of MAO-B inhibitors specifically in managing Parkinson's disease symptoms, examining the evidence for their neuroprotective effects and overall impact on patient quality of life. Understanding how these medications function in clinical settings will further enrich our mastery of dopamine-related therapies.

Chapter 8: MAO–B Inhibitors in Parkinson's Disease

Role in Managing Parkinson's Symptoms

Parkinson's disease (PD) is a progressive neurodegenerative disorder primarily characterized by the loss of dopaminergic neurons in the substantia nigra, leading to a significant reduction in dopamine levels in the striatum. This deficiency results in hallmark motor symptoms, including tremors, rigidity, bradykinesia, and postural instability. MAO-B inhibitors, particularly Selegiline and Rasagiline, have become integral components in the management of Parkinson's disease due to their ability to enhance dopaminergic function.

1. **Mechanism of Action**: By selectively inhibiting MAO-B, these medications prevent the breakdown of dopamine, thereby increasing its availability at the synaptic level. This mechanism addresses the underlying dopamine deficiency in Parkinson's patients, leading to improved motor control and overall symptom management.

2. **Adjunct Therapy**: MAO-B inhibitors are often used as adjuncts to levodopa therapy, which is the primary treatment for Parkinson's disease. They help to smooth out the motor fluctuations associated with levodopa treatment, prolonging its effects and allowing for reduced dosages. This adjunctive use is particularly beneficial as Parkinson's progresses and patients begin to experience "off" periods when their levodopa becomes less effective.

3. **Initial Therapy**: In early-stage Parkinson's disease, MAO-B inhibitors can be prescribed as monotherapy. Clinical studies have shown that both Selegiline and Rasagiline can delay the need for levodopa therapy while providing symptomatic relief, making them effective initial treatment options.

Evidence of Neuroprotective Effects

Beyond their symptomatic benefits, MAO-B inhibitors have been investigated for their potential neuroprotective effects in Parkinson's disease. The oxidative stress associated with dopaminergic degeneration suggests that agents that can reduce this stress may slow disease progression.

1. **Oxidative Stress Reduction**: By inhibiting MAO-B, these medications reduce the formation of harmful reactive oxygen species generated during dopamine metabolism. This reduction in oxidative stress is thought to help protect remaining dopaminergic neurons from further damage.

2. **Clinical Evidence**: Several studies have investigated the neuroprotective effects of Rasagiline and Selegiline. For instance, the ADAGIO trial demonstrated that Rasagiline not only improved motor symptoms but also had the potential to slow disease progression over time. Although the exact mechanisms underlying these neuroprotective effects are still being elucidated, the long-term benefits observed in clinical trials suggest that MAO-B inhibition could have significant implications for the management of Parkinson's disease.

3. **Symptomatic Benefits and Quality of Life**: Patients receiving MAO-B inhibitors often report improved quality of life and reduced disability. Clinical trials indicate that these medications can enhance overall well-being, not just by alleviating motor symptoms but also by positively impacting mood and cognitive function.

Comparative Efficacy of MAO-B Inhibitors

While both Selegiline and Rasagiline are effective MAO-B inhibitors, their comparative efficacy in managing Parkinson's symptoms has been the subject of research and clinical discussion.

1. **Selegiline vs. Rasagiline**: Clinical evidence suggests that both drugs are similarly effective in managing early Parkinson's disease; however, Rasagiline's irreversible binding to MAO-B provides a more sustained effect and may offer better control over motor symptoms in the long term.

2. **Study Outcomes**: In head-to-head studies, Rasagiline has shown comparable efficacy to Selegiline, but with a lower incidence of side effects, particularly those associated with amphetamine metabolites seen in Selegiline. Patients taking Rasagiline may experience fewer stimulant-related side effects, making it a more suitable option for certain populations, such as older adults or those sensitive to stimulants.

3. **Patient Preference and Compliance**: The dosing regimens for both medications also influence patient choice and adherence. Rasagiline's once-daily dosing regimen simplifies treatment, potentially improving patient compliance compared to the more variable dosing required with Selegiline, particularly in its transdermal form.

Conclusion

MAO-B inhibitors, particularly Selegiline and Rasagiline, play a crucial role in the management of Parkinson's disease by enhancing dopamine availability and providing symptomatic relief. Their potential neuroprotective effects add a compelling dimension to their clinical application, offering hope for slowing disease progression and improving quality of life for patients.

In the next chapter, we will explore the impact of MAO-B inhibitors on mental health and depression, examining their potential as adjunct therapies in psychiatric treatment and their effects on mood regulation. This exploration will further highlight the versatility of these medications in addressing dopamine-related disorders beyond Parkinson's disease.

Chapter 9: MAO–B Inhibitors in Mental Health and Depression

Impact on Mood and Potential Antidepressant Benefits

The role of dopamine in mood regulation and emotional well-being has led to increasing interest in the use of MAO-B inhibitors, such as Selegiline and Rasagiline, as adjunctive treatments for depression and related mood disorders. Understanding how these medications interact with dopaminergic pathways provides insight into their potential therapeutic benefits in psychiatric settings.

1. **Dopamine and Mood Disorders**: Dopamine is not only crucial for movement and reward but also significantly influences mood regulation. Low levels of dopamine are associated with depressive symptoms, and alterations in dopaminergic signaling have been implicated in various mood disorders, including major depressive disorder (MDD).

2. **Mechanism of Action in Depression**: MAO-B inhibitors enhance the availability of dopamine by preventing its breakdown, which can lead to improved mood and reduced depressive symptoms. This action is particularly relevant in patients whose depression is characterized by anhedonia (the inability to feel pleasure) or lack of motivation, as increased dopamine levels can promote feelings of reward and pleasure.

3. **Adjunctive Therapy**: MAO-B inhibitors are often considered as adjunct therapies to conventional antidepressants, particularly selective serotonin reuptake inhibitors (SSRIs). Combining these treatments may lead to synergistic effects, enhancing overall therapeutic outcomes for patients with treatment-resistant depression.

Research on MAO-B Inhibitors in Psychiatric Treatment

Research into the effectiveness of MAO-B inhibitors for mood disorders has gained momentum, revealing promising results that support their use in clinical practice.

1. **Clinical Trials**: Several studies have examined the efficacy of Selegiline in treating depression. For example, clinical trials have indicated that Selegiline, especially in its transdermal form, can lead to significant improvements in depressive symptoms with a favorable safety profile. Rasagiline has also been studied for its potential antidepressant effects, showing encouraging results in patients with comorbid Parkinson's disease and depression.

2. **Efficacy in Specific Populations**: MAO-B inhibitors may be particularly beneficial for patients who do not respond adequately to traditional antidepressant therapies. In individuals with Parkinson's disease, where depression is common, the use of Rasagiline has demonstrated improvements not only in motor symptoms but also in mood, thereby addressing both aspects of the disease.

3. **Safety Profile**: The safety profile of MAO-B inhibitors is generally favorable, especially compared to non-selective MAO inhibitors, which can require dietary restrictions to avoid hypertensive crises. The selectivity of MAO-B inhibitors minimizes these risks, allowing for a broader patient population to benefit from their use.

Evidence Supporting Use in Depression

The accumulating evidence supporting the use of MAO-B inhibitors in treating mood disorders underscores their potential as valuable tools in psychiatric practice.

1. **Mechanistic Insights**: The neuroprotective properties of MAO-B inhibitors may contribute to their antidepressant effects. By reducing oxidative stress and promoting neuronal health, these drugs can enhance overall brain function, which is critical for mood regulation.

2. **Case Studies and Observational Data**: Real-world applications and case studies have illustrated the effectiveness of MAO-B inhibitors in patients with depression. Many patients report improved mood, increased energy, and better engagement in daily activities after initiating treatment with Selegiline or Rasagiline.

3. **Long-Term Outcomes**: Long-term studies are beginning to suggest that the use of MAO-B inhibitors may lead to sustained improvements in mood and cognition, making them an attractive option for chronic mood disorders. Ongoing research will be essential to further elucidate their role in this area.

Potential Side Effects and Risks

While MAO-B inhibitors offer potential benefits in treating depression, it is important to be aware of the potential side effects and risks associated with their use.

1. **Common Side Effects**: Patients may experience side effects such as insomnia, dizziness, and gastrointestinal disturbances. Although these are generally mild, they can impact treatment adherence.

2. **Interactions with Other Medications**: Caution is advised when combining MAO-B inhibitors with other antidepressants, especially SSRIs and other medications that increase serotonin levels, due to the risk of serotonin syndrome. Proper screening and monitoring are critical when initiating treatment with MAO-B inhibitors in patients already on antidepressant therapy.

3. **Dietary Considerations**: Although MAO-B inhibitors carry a lower risk of dietary interactions compared to non-selective MAO inhibitors, patients should still be educated about potential dietary restrictions, particularly regarding tyramine-containing foods, especially if high doses are used.

Conclusion

The use of MAO-B inhibitors such as Selegiline and Rasagiline in managing mood disorders represents a promising avenue in psychiatric treatment. By enhancing dopamine availability and potentially offering neuroprotective benefits, these medications can improve mood and overall quality of life for individuals suffering from depression, particularly in complex cases where traditional therapies may fall short.

In the next chapter, we will delve into the intricate workings of the brain's dopaminergic pathways and how MAO-B inhibitors influence these mechanisms. This understanding will further illuminate the broader implications of dopamine modulation in both neurodegenerative and psychiatric disorders, providing a holistic view of their therapeutic potential.

Chapter 10: Dopaminergic Pathways and Mechanisms

Overview of Dopaminergic Pathways

Dopamine is a pivotal neurotransmitter in the central nervous system, influencing a wide array of physiological and psychological processes. Understanding the brain's dopaminergic pathways is essential for appreciating how MAO-B inhibitors like Selegiline and Rasagiline exert their effects. Dopaminergic pathways are primarily classified into four major circuits, each serving distinct functions in mood regulation, motor control, cognition, and reward processing.

1. **Nigrostriatal Pathway**: This pathway connects the substantia nigra to the striatum and is crucial for the regulation of voluntary motor control. It is primarily affected in Parkinson's disease, where degeneration of dopaminergic neurons leads to the characteristic motor symptoms.

2. **Mesolimbic Pathway**: Originating from the ventral tegmental area (VTA), this pathway projects to the nucleus accumbens and is heavily involved in reward processing, motivation, and the reinforcement of pleasurable experiences. Dysregulation of this pathway is implicated in addiction and mood disorders.

3. **Mesocortical Pathway**: Also arising from the VTA, this pathway projects to the prefrontal cortex and is associated with higher cognitive functions, including decision-making, attention, and impulse control. Alterations in this pathway can contribute to cognitive deficits in schizophrenia and other psychiatric disorders.

4. **Tuberoinfundibular Pathway**: This pathway connects the hypothalamus to the pituitary gland and regulates the secretion of prolactin. Dopamine acts as an inhibitor in this pathway, demonstrating the diverse roles of dopamine in endocrine regulation.

How MAO-B Inhibitors Influence Dopaminergic Pathways

MAO-B inhibitors play a critical role in enhancing dopaminergic transmission by increasing dopamine availability in the brain. The mechanisms through which Selegiline and Rasagiline exert their effects on these pathways are multifaceted.

1. **Inhibition of Dopamine Breakdown**: By selectively inhibiting MAO-B, these medications prevent the metabolic breakdown of dopamine, leading to increased dopamine levels in the synaptic cleft. This action is particularly beneficial in the nigrostriatal pathway, where dopamine deficiency is a hallmark of Parkinson's disease.

2. **Enhancement of Dopaminergic Signaling**: Increased dopamine levels enhance signaling through dopamine receptors, improving motor function and mood regulation. In the mesolimbic and mesocortical pathways, elevated dopamine may also contribute to enhanced motivation and cognitive function.

3. **Neuroprotective Mechanisms**: The reduction of oxidative stress through MAO-B inhibition not only preserves dopamine levels but also protects dopaminergic neurons from damage. This neuroprotective effect is particularly relevant in the context of neurodegenerative diseases, where oxidative stress contributes to neuronal loss.

4. **Modulation of Other Neurotransmitter Systems**: MAO-B inhibitors may also have indirect effects on other neurotransmitter systems, such as serotonin and norepinephrine. By stabilizing dopamine levels, these medications could enhance the overall balance of neurotransmitter activity in the brain, contributing to their antidepressant effects.

Dopaminergic Pathway Interactions

The interaction between dopaminergic pathways is crucial for understanding the comprehensive effects of MAO-B inhibitors.

1. **Dopamine and Serotonin Interactions**: The interplay between dopamine and serotonin systems is significant, particularly in mood regulation. MAO-B inhibitors may enhance dopaminergic signaling, which can positively influence serotonergic activity, thereby contributing to improvements in mood and alleviation of depressive symptoms.

2. **Reward Pathways and Addiction**: The mesolimbic pathway's role in reward processing highlights how MAO-B inhibitors might influence behaviors associated with addiction. By enhancing dopamine availability in this pathway, these medications could potentially support recovery efforts in individuals with substance use disorders.

3. **Cognitive Function and the Prefrontal Cortex**: The mesocortical pathway's relationship with cognitive function underscores the importance of maintaining optimal dopamine levels. MAO-B inhibitors may enhance cognitive function by stabilizing dopaminergic activity in the prefrontal cortex, which is crucial for executive function and decision-making.

Implications for Treatment

Understanding the mechanisms through which MAO-B inhibitors influence dopaminergic pathways has important implications for their use in clinical practice.

1. **Personalized Treatment Strategies**: Clinicians can tailor treatment approaches based on a patient's specific symptoms and the underlying dopaminergic pathways involved. For instance, patients with prominent motor symptoms may benefit more from MAO-B inhibitors as adjuncts to levodopa therapy, while those with mood disorders may find these medications effective in improving their overall well-being.

2. **Combination Therapies**: The synergistic effects of MAO-B inhibitors when combined with other dopaminergic treatments, such as levodopa or dopamine agonists, can optimize therapeutic outcomes. Understanding the interactions between these pathways allows for more effective management of complex cases of Parkinson's disease and associated mood disorders.

3. **Future Research Directions**: Continued research into the specific effects of MAO-B inhibitors on dopaminergic pathways may uncover additional therapeutic applications, particularly in conditions like schizophrenia, depression, and cognitive decline associated with aging.

Conclusion

The intricate relationships among the brain's dopaminergic pathways highlight the significant role that MAO-B inhibitors like Selegiline and Rasagiline play in modulating dopamine levels and enhancing neurological function. By understanding these mechanisms, healthcare professionals can make informed decisions about treatment strategies, optimizing outcomes for patients with Parkinson's disease and mood disorders.

In the following chapter, we will examine potential side effects and risks associated with MAO-B inhibitors, including strategies for effective management and mitigation of these effects. This knowledge will further inform clinical practice and ensure safer, more effective use of these valuable therapeutic agents.

Chapter 11: Potential Side Effects and Risks

As with any pharmacological intervention, the use of MAO-B inhibitors like Selegiline and Rasagiline carries potential side effects and risks that clinicians and patients must carefully consider. Understanding these aspects is crucial for effective management and optimizing therapeutic outcomes while minimizing adverse effects.

Common Side Effects of Selegiline and Rasagiline

Both Selegiline and Rasagiline are generally well-tolerated; however, they can produce a range of side effects. Awareness of these potential adverse reactions enables healthcare providers to monitor patients closely and intervene when necessary.

1. **Gastrointestinal Disturbances:**

   Patients may experience nausea, vomiting, diarrhea, or constipation. These symptoms are often mild and can sometimes be mitigated by adjusting the dosage or taking the medication with food.

2. **Central Nervous System Effects:**

   - **Insomnia**: Some patients may experience sleep disturbances, including insomnia or difficulty falling asleep. This effect is particularly associated with Selegiline, likely due to its stimulant metabolites.
   - **Dizziness and Lightheadedness**: These symptoms can occur, especially during the initial phase of treatment or with dose increases.

3. **Headache:**

   Headaches are a common side effect and may be related to changes in neurotransmitter levels.

4. **Orthostatic Hypotension:**

Some patients may experience a drop in blood pressure upon standing, leading to dizziness or fainting. This side effect is more common in patients taking multiple medications or those with pre-existing cardiovascular conditions.

5. **Fatigue and Lethargy:**

While these medications are often intended to enhance energy levels, some individuals may experience fatigue or feelings of lethargy, particularly when first starting treatment.

6. **Impulse Control Disorders:**

There have been reports of increased impulsivity or compulsive behaviors in some patients, particularly in those with a history of behavioral issues.

Precautions, Contraindications, and Risk Mitigation

While MAO-B inhibitors can be highly effective, certain precautions and contraindications must be taken into account to minimize risks.

1. **Dietary Restrictions:**

Although MAO-B inhibitors are generally less restrictive than non-selective MAO inhibitors, patients should still be advised to avoid foods high in tyramine, especially when taking higher doses. Tyramine-rich foods include aged cheeses, cured meats, fermented products, and certain alcoholic beverages. High tyramine intake can lead to hypertensive crises, although the risk is lower with MAO-B inhibitors.

2. **Drug Interactions:**

Caution is necessary when prescribing MAO-B inhibitors alongside other medications, particularly those that affect serotonin levels, such as SSRIs or SNRIs. The combination can increase the risk of serotonin syndrome, a potentially life-threatening condition characterized by symptoms such as confusion, agitation, rapid heart rate, and increased blood pressure.

3. **Caution in Patients with Comorbid Conditions:**

Patients with a history of cardiovascular disease, hepatic impairment, or those who are elderly may require careful monitoring and dose adjustments to mitigate risks.

4. **Monitoring for Side Effects:**

Regular follow-up appointments are essential for monitoring side effects and ensuring that patients are adhering to dietary guidelines. Clinicians should be vigilant for signs of impulse control disorders, particularly in patients with a history of behavioral problems.

Long-Term Use and Tolerance

The long-term use of MAO-B inhibitors presents its own set of considerations, particularly regarding tolerance and the sustainability of their effects.

1. **Tolerance Development:**

   Some patients may develop tolerance to the effects of MAO-B inhibitors over time, leading to reduced efficacy in managing symptoms. This phenomenon necessitates ongoing evaluation of treatment effectiveness and possible adjustments in therapy.

2. **Assessment of Long-Term Efficacy:**

   Long-term studies have shown that while MAO-B inhibitors can provide sustained benefits for managing Parkinson's disease, ongoing assessment is crucial to determine whether treatment should be continued or adjusted. Clinicians should periodically reevaluate the necessity of MAO-B inhibitors in conjunction with other treatments.

3. **Strategies for Managing Tolerance:**

   If tolerance develops, strategies such as dose adjustments, drug holidays, or switching to alternative therapies may be explored to regain therapeutic efficacy.

Conclusion

Understanding the potential side effects and risks associated with MAO-B inhibitors like Selegiline and Rasagiline is essential for their safe and effective use. By being aware of common adverse reactions, contraindications, and the importance of monitoring, clinicians can optimize treatment strategies while minimizing risks for patients.

In the following chapter, we will explore interactions with other medications, particularly those that may impact serotonin levels, and discuss the importance of careful management in conjunction with MAO-B inhibitors. This will further enhance our understanding of how to effectively integrate these medications into comprehensive treatment plans for patients with dopamine-related disorders.

Chapter 12: Interactions with Other Medications

Importance of Careful Management with Other Therapies

When treating patients with MAO-B inhibitors such as Selegiline and Rasagiline, it is essential to consider potential drug interactions. These interactions can significantly influence the efficacy and safety of treatment, highlighting the need for careful management and monitoring. Understanding how these medications interact with other classes of drugs allows clinicians to make informed decisions, ensuring optimal therapeutic outcomes.

Common Drug Interactions

1. **Selective Serotonin Reuptake Inhibitors (SSRIs):**

MAO-B inhibitors can interact with SSRIs, leading to an increased risk of serotonin syndrome. This potentially life-threatening condition is characterized by symptoms such as confusion, agitation, rapid heart rate, and elevated blood pressure. Symptoms can escalate to severe complications, including hyperthermia and seizures. Therefore, it is crucial to monitor patients closely if an MAO-B inhibitor is prescribed alongside SSRIs.

2. **Other Antidepressants:**

In addition to SSRIs, other antidepressants, particularly those that affect serotonin levels (such as SNRIs and tricyclic antidepressants), may also pose risks when used in conjunction with MAO-B inhibitors. The combination can lead to excessive serotonin levels, and similar precautions should be observed as with SSRIs.

3. **Dopamine Agonists:**

The concurrent use of MAO-B inhibitors with dopamine agonists (e.g., pramipexole and ropinirole) is common in the treatment of Parkinson's disease. While these combinations can enhance dopaminergic effects and improve symptom management, clinicians should remain vigilant for signs of excessive dopaminergic stimulation, including hallucinations and dyskinesias.

4. **Sympathomimetic Agents:**

Caution is warranted when combining MAO-B inhibitors with sympathomimetic drugs, such as decongestants (e.g., pseudoephedrine) and certain medications for attention-deficit/hyperactivity disorder (ADHD) that may increase norepinephrine and epinephrine levels. This combination can potentially lead to hypertensive crises or other cardiovascular complications due to elevated catecholamines.

5. **CYP450 Inhibitors and Inducers:**

As Selegiline and Rasagiline are metabolized by the cytochrome P450 system, particularly CYP1A2, the concurrent use of drugs that inhibit or induce these enzymes can significantly affect the metabolism of MAO-B inhibitors. For instance, medications like fluvoxamine (an SSRI) can inhibit CYP1A2, potentially increasing the plasma concentration of Rasagiline and increasing the risk of side effects. Conversely, inducers can lower plasma levels, diminishing the efficacy of MAO-B inhibitors.

Management Strategies

1. **Comprehensive Medication Review:**

   Clinicians should conduct thorough medication reviews for patients prescribed MAO-B inhibitors, assessing for potential interactions with all concurrent medications, including over-the-counter products and supplements.

2. **Patient Education:**

   Educating patients about potential interactions and the signs of serotonin syndrome or hypertensive crises is vital. Patients should be informed about which medications to avoid and the importance of disclosing all medications they are taking, including herbal supplements.

3. **Monitoring and Adjustments:**

Close monitoring of patients on MAO-B inhibitors is essential, particularly during the initiation of therapy or when adding other medications. Adjustments in dosages may be necessary based on patient response and any adverse effects experienced.

4. **Alternative Therapies:**

When possible, exploring alternative therapies that do not pose the same interaction risks can enhance patient safety. For instance, using non-interacting antidepressants or adjusting the dosage of existing medications may mitigate potential complications.

Conclusion

The interactions of MAO-B inhibitors with other medications underscore the importance of careful management in clinical practice. By understanding these interactions and implementing appropriate monitoring and patient education strategies, clinicians can optimize treatment outcomes while minimizing risks.

In the next chapter, we will discuss long-term efficacy and tolerance, exploring how the extended use of MAO-B inhibitors affects patient outcomes and strategies for managing potential tolerance development. This will further enhance our understanding of the role of MAO-B inhibitors in chronic dopamine-related therapies.

Chapter 13: Long–Term Efficacy and Tolerance

Insights into Tolerance Development

As with many pharmacological treatments, the long-term use of MAO-B inhibitors like Selegiline and Rasagiline raises concerns about the development of tolerance. Tolerance occurs when the response to a drug diminishes over time, necessitating higher doses to achieve the same therapeutic effect. Understanding the mechanisms and implications of tolerance in the context of MAO-B inhibitors is crucial for optimizing treatment strategies and improving patient outcomes.

1. **Mechanisms of Tolerance:**

   - **Receptor Adaptation**: One of the primary mechanisms of tolerance may involve adaptive changes in dopamine receptors. Chronic elevation of dopamine levels due to MAO-B inhibition can lead to receptor downregulation or desensitization, reducing the overall responsiveness to dopamine signaling.

   - **Homeostatic Adjustments**: The brain has complex feedback mechanisms that strive to maintain homeostasis. Prolonged MAO-B inhibition may trigger compensatory changes in other neurotransmitter systems, such as serotonin or norepinephrine, which can further complicate the pharmacological effects of MAO-B inhibitors.

   - **Neuronal Plasticity**: Chronic treatment with MAO-B inhibitors may influence neuronal plasticity, leading to alterations in synaptic connections and neurotransmitter release. These changes can affect the overall dopaminergic signaling pathways and may result in diminished therapeutic efficacy over time.

2. **Clinical Implications of Tolerance:**

- **Diminished Efficacy**: Patients may experience a reduction in the effectiveness of MAO-B inhibitors in managing their symptoms, particularly in terms of motor function and mood stabilization. This can lead to the necessity of increasing the dose, which may not always be advisable due to the risk of side effects and potential drug interactions.

- **Adjustment of Treatment Plans**: Clinicians should be vigilant in monitoring for signs of tolerance and be prepared to adjust treatment plans accordingly. This may include switching medications, adding adjunct therapies, or implementing drug holidays where appropriate.

Managing and Understanding Long-Term Effects

Effectively managing the long-term use of MAO-B inhibitors involves several strategies to address potential tolerance and ensure sustained therapeutic benefits.

1. **Regular Monitoring:**

Regular follow-ups and assessments of patient symptoms are essential for identifying any signs of tolerance. Clinicians should use standardized rating scales to evaluate motor function, mood, and overall quality of life, allowing for objective measurements of treatment efficacy.

2. **Dose Adjustments:**

Clinicians may need to adjust the dosing regimen based on the patient's response over time. In some cases, a lower dose may suffice if tolerance has developed, while in others, a slight increase may be warranted to maintain symptom control. Close attention to side effects is critical during these adjustments.

3. **Combination Therapies:**

The use of adjunctive therapies, such as dopamine agonists or other classes of antidepressants, may enhance the overall treatment effect and help mitigate tolerance. Combining therapies allows for a more comprehensive approach to symptom management, addressing multiple aspects of Parkinson's disease or depressive disorders.

4. **Drug Holidays:**

In certain cases, clinicians may consider implementing a drug holiday, temporarily discontinuing the MAO-B inhibitor to allow receptor sensitivity to recover. This strategy should be approached cautiously and requires careful planning to minimize the risk of symptom exacerbation.

5. **Patient Education:**

Educating patients about the potential for tolerance and the importance of adhering to prescribed treatment regimens is vital. Patients should be informed about what to expect during long-term therapy, including the potential need for adjustments and the importance of communicating changes in their symptoms to their healthcare providers.

Future Research Directions

Research into the long-term efficacy and tolerance associated with MAO-B inhibitors is ongoing, with several areas of focus that may enhance our understanding and management of these medications.

1. **Mechanistic Studies:**

   Further investigations into the mechanisms underlying tolerance development will provide insights into how the brain adapts to prolonged MAO-B inhibition. Understanding these pathways can inform strategies to mitigate tolerance and enhance therapeutic outcomes.

2. **Longitudinal Clinical Trials:**

   Long-term clinical trials examining the efficacy and safety of MAO-B inhibitors over extended periods will help elucidate the sustainability of their benefits. Such studies can clarify the role of these medications in chronic management scenarios.

3. **Alternative Delivery Methods:**

   Exploring different delivery methods, such as continuous infusion or sustained-release formulations, may help maintain stable plasma levels of MAO-B inhibitors and reduce fluctuations that could contribute to tolerance.

4. **Combination Studies:**

Investigating the efficacy of MAO-B inhibitors in combination with novel therapies or agents targeting different neurotransmitter systems could yield promising results in managing tolerance and enhancing overall treatment effectiveness.

Conclusion

Understanding long-term efficacy and tolerance in MAO-B inhibitors is essential for optimizing their use in clinical practice. By monitoring patients closely, adjusting treatment plans as needed, and remaining vigilant for signs of tolerance, healthcare providers can ensure that these medications continue to provide benefits over time.

In the next chapter, we will explore the application of Selegiline in other neurological conditions, examining current research into its efficacy beyond Parkinson's disease and potential off-label uses in treating cognitive disorders. This exploration will highlight the versatility and potential of MAO-B inhibitors in a broader therapeutic context.

Chapter 14: Selegiline in Other Neurological Conditions

While Selegiline is primarily recognized for its role in managing Parkinson's disease, its potential therapeutic applications extend beyond this condition. Research into Selegiline's effects on various neurological disorders has generated interest in its use for cognitive decline, Alzheimer's disease, and other conditions involving dopaminergic dysfunction. This chapter explores the current understanding and emerging research regarding Selegiline's applications in these areas.

Research into Alzheimer's Disease and Cognitive Disorders

Alzheimer's disease (AD) is the most common form of dementia, characterized by progressive cognitive decline, memory loss, and behavioral changes. Given the dopaminergic deficits observed in some patients with AD, researchers have investigated the use of Selegiline as a potential adjunct therapy.

1. **Mechanism of Action:**

In Alzheimer's disease, there is a significant loss of cholinergic neurons, which contributes to cognitive deficits. However, dopaminergic systems are also affected. Selegiline's ability to increase dopamine levels could provide some symptomatic relief by enhancing dopaminergic signaling in the brain, which may have positive implications for cognition and behavior.

2. **Clinical Studies:**

Several small-scale studies have explored the efficacy of Selegiline in patients with Alzheimer's disease. While results have been mixed, some studies reported modest improvements in cognitive function and behavior, particularly in patients with early-stage dementia. The neuroprotective properties of Selegiline, owing to its antioxidant effects, may also contribute to its potential benefits in AD.

3. **Combination Therapy:**

The possibility of using Selegiline in conjunction with cholinesterase inhibitors (such as donepezil) is of particular interest. Combining these therapies may address both cholinergic and dopaminergic deficits, potentially leading to improved cognitive outcomes. However, more extensive and well-designed clinical trials are necessary to establish the efficacy and safety of such combinations.

Selegiline in Other Neurodegenerative Conditions

In addition to Alzheimer's disease, Selegiline has been investigated for its potential benefits in various neurodegenerative conditions characterized by dopaminergic dysfunction.

1. **Huntington's Disease:**

Huntington's disease (HD) is a hereditary neurodegenerative disorder that leads to motor dysfunction, cognitive decline, and psychiatric symptoms. Research has indicated that Selegiline may help manage some symptoms in HD patients by enhancing dopaminergic transmission and potentially providing neuroprotective effects. Some clinical observations suggest improvements in mood and behavioral symptoms, although robust clinical evidence is still lacking.

2. **Multiple Sclerosis:**

Multiple sclerosis (MS) is a demyelinating disease that can affect motor and cognitive function. While the primary treatment strategies focus on immunomodulation, some studies have explored the use of Selegiline for managing fatigue and cognitive symptoms in MS patients. Preliminary findings suggest potential benefits, but further research is needed to substantiate these observations.

3. **Cognitive Impairment in Aging:**

Cognitive impairment associated with aging may also benefit from Selegiline therapy. Some studies suggest that Selegiline may help improve cognitive function in elderly patients, particularly in those experiencing age-related cognitive decline. The underlying mechanism may involve enhanced dopaminergic activity, which is essential for cognitive processes such as attention, memory, and executive function.

Experimental and Off-Label Uses

The flexibility of Selegiline's mechanism of action has led to interest in its off-label use for various conditions beyond traditional indications.

1. **Chronic Pain Management:**

   Some preliminary research has suggested that Selegiline may have a role in managing chronic pain, particularly neuropathic pain. Its influence on dopaminergic pathways might provide a novel approach to treating pain that is resistant to conventional therapies.

2. **Addiction and Substance Use Disorders:**

   Given its effects on the mesolimbic dopamine pathway, there is interest in exploring Selegiline's potential in treating addiction, particularly in reducing cravings and withdrawal symptoms in patients with substance use disorders.

3. **Potential as a Nootropic:**

The cognitive-enhancing properties of Selegiline have led to investigations into its use as a nootropic agent. Some studies suggest that it may improve memory, learning, and cognitive performance in healthy individuals, though more research is needed to substantiate these claims.

Conclusion

Selegiline's potential applications beyond Parkinson's disease highlight its versatility as a therapeutic agent. While evidence supporting its use in conditions such as Alzheimer's disease, Huntington's disease, and other neurodegenerative disorders is emerging, further research is crucial to fully understand its efficacy and safety in these contexts.

In the next chapter, we will explore Rasagiline and its future in neuroprotection, discussing ongoing research directions and potential applications beyond its established role in Parkinson's disease. This exploration will provide a comprehensive view of the evolving landscape of MAO-B inhibitors and their significance in neuropharmacology.

Chapter 15: Rasagiline and Its Future in Neuroprotection

Introduction to Rasagiline

Rasagiline, a selective and irreversible monoamine oxidase B (MAO-B) inhibitor, has gained prominence in the treatment of Parkinson's disease due to its ability to enhance dopaminergic signaling and provide neuroprotective benefits. Understanding Rasagiline's mechanisms, current applications, and future research directions is critical for recognizing its potential beyond managing Parkinson's disease.

Mechanism of Action and Neuroprotective Properties

1. **Selective MAO-B Inhibition:**

Rasagiline selectively inhibits the MAO-B enzyme, which is responsible for the breakdown of dopamine in the brain. This inhibition results in increased dopamine availability, particularly beneficial in the treatment of Parkinson's disease.

2. **Neuroprotection:**

Beyond its dopaminergic effects, Rasagiline exhibits neuroprotective properties. Studies suggest that it can reduce oxidative stress and inflammation, which are critical factors in the progression of neurodegenerative diseases. By minimizing oxidative damage to neurons, Rasagiline may help preserve neuronal integrity and function.

3. **Anti-inflammatory Effects:**

Research indicates that Rasagiline may exert anti-inflammatory effects, which can further support neuronal health. By reducing neuroinflammation, Rasagiline may mitigate the secondary damage associated with dopaminergic neuron loss.

Current Applications in Parkinson's Disease

Rasagiline is primarily used in managing Parkinson's disease, both as monotherapy in early stages and as an adjunct to levodopa therapy in advanced cases. Its role in improving motor symptoms and overall quality of life is well established.

1. **Symptomatic Relief:**

   Clinical trials have demonstrated that Rasagiline effectively improves motor symptoms and may delay the progression of the disease when used early in the treatment regimen. Patients often experience improved motor function, reduced "off" periods, and enhanced quality of life.

2. **Long-term Benefits:**

   The neuroprotective effects of Rasagiline are particularly valuable as they may slow disease progression, a critical consideration in managing chronic neurodegenerative disorders. Ongoing studies continue to explore the extent and duration of these benefits.

Future Research Directions

As the understanding of Rasagiline's mechanisms and effects deepens, several exciting research avenues are emerging.

1. **Investigating Efficacy in Other Neurodegenerative Disorders:**

Given its neuroprotective properties, Rasagiline is being explored for potential applications beyond Parkinson's disease. Research is underway to assess its efficacy in conditions such as Alzheimer's disease, Huntington's disease, and multiple sclerosis. The hypothesis is that Rasagiline's ability to enhance dopaminergic activity and reduce oxidative stress may also confer benefits in these disorders.

2. **Combination Therapies:**

Future studies will likely investigate the effectiveness of Rasagiline in combination with other therapeutic agents, such as cholinesterase inhibitors in Alzheimer's disease or other neuroprotective agents. These combinations could provide synergistic effects and improve overall treatment outcomes.

3. **Exploring the Mechanisms of Neuroprotection:**

   Further research into the specific neuroprotective mechanisms of Rasagiline is essential. Understanding how it modulates cellular pathways involved in oxidative stress, inflammation, and neurodegeneration could reveal new therapeutic targets and strategies.

4. **Long-term Safety and Efficacy Studies:**

   Longitudinal studies are needed to evaluate the long-term safety and efficacy of Rasagiline, particularly as patients transition through different stages of Parkinson's disease and related disorders. Monitoring for potential side effects and assessing quality of life outcomes over extended periods will be crucial.

Conclusion

Rasagiline represents a significant advancement in the management of Parkinson's disease and holds promise for broader applications in neuroprotection. Its ability to enhance dopaminergic signaling while providing neuroprotective benefits positions it as a valuable therapeutic agent in the landscape of neurodegenerative disorders.

As research continues to unfold, Rasagiline's potential applications beyond Parkinson's disease could redefine its role in clinical practice. The ongoing exploration of its mechanisms, efficacy in other conditions, and long-term outcomes will enhance our understanding of dopaminergic therapies and their impact on neurodegeneration.

In the next chapter, we will discuss dopamine and lifestyle factors, examining how diet, exercise, and other lifestyle choices can influence dopamine levels and overall health. This exploration will provide insights into holistic practices that can complement pharmacological treatments and optimize dopaminergic health.

Chapter 16: Dopamine and Lifestyle Factors

Introduction

Dopamine plays a crucial role in numerous physiological and psychological functions, making it essential for maintaining overall health and well-being. Recent research has highlighted the impact of lifestyle factors such as diet, exercise, and stress management on dopamine levels and function. This chapter explores how these lifestyle choices can synergistically interact with MAO-B inhibitors like Selegiline and Rasagiline to optimize dopaminergic health.

Diet and Dopamine Levels

1. **Nutritional Sources:**

   - **Amino Acids**: The primary building blocks for dopamine synthesis are the amino acids phenylalanine and tyrosine. Foods rich in these amino acids include lean proteins (e.g., chicken, turkey, fish), dairy products, nuts, seeds, and legumes. Incorporating these foods into the diet can support dopamine production.

   - **Antioxidants**: Oxidative stress is detrimental to dopamine-producing neurons. Antioxidant-rich foods such as berries, dark chocolate, green leafy vegetables, and nuts help combat oxidative damage, potentially enhancing dopamine health.

2. **Omega-3 Fatty Acids:**

Research suggests that omega-3 fatty acids, found in fatty fish (e.g., salmon, mackerel), flaxseeds, and walnuts, can improve dopamine receptor sensitivity. This enhances the efficacy of dopamine signaling and may offer additional neuroprotective benefits.

3. **Gut Health:**

The gut-brain axis is increasingly recognized for its role in neurotransmitter regulation, including dopamine. A diet rich in fiber, fermented foods, and probiotics can support gut health and promote the production of gut-derived neurotransmitters, positively influencing overall dopamine levels.

4. **Dietary Patterns:**

Diets that are high in saturated fats and sugars may negatively impact dopamine signaling and receptor function. A balanced diet that emphasizes whole foods, healthy fats, and adequate protein can support optimal dopaminergic health.

Exercise and Dopamine

1. **Physical Activity:**

Regular physical activity has been shown to boost dopamine levels and enhance receptor sensitivity. Exercise stimulates the release of various neurotransmitters, including dopamine, leading to improved mood, motivation, and cognitive function.

2. **Aerobic Exercise:**

Aerobic activities such as running, cycling, and swimming are particularly effective in promoting dopamine release. Research has shown that engaging in moderate to vigorous aerobic exercise can increase dopamine levels and improve overall brain health.

3. **Resistance Training:**

Strength training has also been associated with improved dopaminergic function. Studies indicate that engaging in resistance training can enhance mood and cognitive performance, potentially through its effects on neurotransmitter systems.

4. **Exercise as a Complement to Pharmacotherapy:**

For patients taking MAO-B inhibitors, regular exercise may enhance the therapeutic effects of these medications. By improving overall dopamine signaling, exercise can work synergistically with Selegiline and Rasagiline to optimize treatment outcomes.

Stress Management

1. **Impact of Stress on Dopamine:**

   Chronic stress can lead to dysregulation of dopamine systems, contributing to mood disorders and cognitive decline. Managing stress is essential for maintaining healthy dopamine levels and supporting overall mental health.

2. **Mindfulness and Relaxation Techniques:**

   Practices such as mindfulness meditation, yoga, and deep-breathing exercises have been shown to reduce stress and promote emotional well-being. These practices can enhance dopamine function and receptor sensitivity, leading to improved mood and resilience.

3. **Social Support:**

   Social interactions and support systems play a significant role in managing stress. Positive social connections can enhance dopamine release, contributing to feelings of happiness and fulfillment.

Synergistic Effects with MAO-B Inhibitors

1. **Holistic Approach to Treatment:**

   Integrating lifestyle factors with pharmacological treatments can enhance overall therapeutic efficacy. MAO-B inhibitors like Selegiline and Rasagiline may work more effectively in individuals who maintain a healthy lifestyle, characterized by a balanced diet, regular exercise, and effective stress management.

2. **Patient-Centered Care:**

   Clinicians should encourage patients to adopt lifestyle modifications alongside their medication regimen. This approach not only addresses dopaminergic health but also promotes overall physical and mental well-being.

3. **Case Examples:**

Illustrative case studies can highlight the positive outcomes associated with combining lifestyle factors with MAO-B inhibitors. For example, a patient who engages in regular exercise while on Rasagiline may report improved mood and reduced motor fluctuations compared to a sedentary patient on the same medication.

Conclusion

Dopamine's influence on health extends beyond pharmacological interventions, and lifestyle factors play a crucial role in optimizing dopaminergic function. By focusing on a balanced diet, regular physical activity, and effective stress management, patients can enhance the benefits of MAO-B inhibitors like Selegiline and Rasagiline, ultimately improving their quality of life.

In the next chapter, we will explore alternative dopaminergic therapies, examining other treatment options available for managing conditions associated with dopamine dysregulation. This exploration will provide a comprehensive view of the diverse therapeutic landscape for dopamine-related disorders.

Chapter 17: Alternative Dopaminergic Therapies

Overview of Alternative Treatments

While MAO-B inhibitors such as Selegiline and Rasagiline play a crucial role in the management of Parkinson's disease and other dopamine-related disorders, several alternative therapies can also be considered to enhance dopaminergic function and address the multifaceted nature of these conditions. This chapter explores various alternative dopaminergic therapies, including pharmacological options, natural supplements, and non-pharmacological interventions.

Pharmacological Alternatives

1. **Levodopa:**

   Levodopa is the most effective treatment for Parkinson's disease and serves as a cornerstone of dopaminergic therapy. It is a precursor to dopamine that can cross the blood-brain barrier and be converted into dopamine, replenishing depleted levels in the brain. While highly effective, long-term use can lead to motor fluctuations and dyskinesias.

2. **Dopamine Agonists:**

   Medications such as pramipexole and ropinirole act directly on dopamine receptors, mimicking the effects of dopamine in the brain. They can be used as monotherapy in early-stage Parkinson's disease or as adjuncts to levodopa in advanced stages. Dopamine agonists are associated with a lower risk of motor complications than levodopa but may cause side effects such as impulse control disorders.

3. **Amantadine:**

Originally developed as an antiviral medication, amantadine has been found to have dopaminergic effects, particularly in alleviating dyskinesias associated with levodopa treatment. Its mechanism is believed to involve the release of dopamine and antagonism of NMDA receptors, contributing to its therapeutic benefits in Parkinson's disease.

4. **Anticholinergics:**

Medications such as trihexyphenidyl and benztropine can help manage tremors and rigidity, particularly in younger patients with Parkinson's disease. These agents work by blocking the effects of acetylcholine, which can become dysregulated in the absence of adequate dopamine.

5. **Catechol-O-Methyltransferase (COMT) Inhibitors:**

Agents like entacapone and tolcapone inhibit the COMT enzyme, which breaks down levodopa. By prolonging the effects of levodopa, these medications can help reduce "wearing-off" phenomena and improve overall motor control when used in conjunction with levodopa therapy.

Natural Supplements

1. **Tyrosine and Phenylalanine:**

These amino acids are precursors to dopamine synthesis. Supplementation with tyrosine or phenylalanine may support dopamine production, especially during periods of stress or when dietary intake is insufficient.

2. **Omega-3 Fatty Acids:**

Omega-3s, found in fish oil and flaxseed oil, have been shown to promote dopaminergic function and protect against neurodegeneration. Their anti-inflammatory properties may also support overall brain health.

3. **Ginkgo Biloba:**

This herbal supplement is thought to improve cognitive function and may enhance dopamine signaling. Some studies have suggested potential benefits for cognitive decline, although more research is needed to confirm its efficacy.

4. **Rhodiola Rosea:**

Known for its adaptogenic properties, Rhodiola may enhance dopamine levels and improve mood and cognitive function, particularly under stress. Preliminary research indicates potential benefits in reducing fatigue and enhancing mental performance.

Non-Pharmacological Interventions

1. **Exercise:**

   Regular physical activity has been consistently associated with increased dopamine levels and improved mood. Exercise can stimulate the release of dopamine and promote neuroplasticity, which may help offset some symptoms of Parkinson's disease and cognitive decline.

2. **Cognitive Behavioral Therapy (CBT):**

   For patients experiencing mood disorders related to dopamine dysregulation, CBT and other forms of psychotherapy can provide essential support. These therapies can help patients develop coping strategies, enhance motivation, and improve overall well-being.

3. **Mindfulness and Meditation:**

Practices such as mindfulness meditation have been shown to reduce stress and enhance mood, potentially influencing dopamine release and receptor sensitivity. Mindfulness can also improve overall quality of life for patients with chronic conditions.

4. **Dietary Modifications:**

As previously discussed, a diet rich in antioxidants, healthy fats, and essential amino acids can support dopaminergic health. Making conscious dietary choices can enhance the effects of pharmacological treatments and improve overall brain function.

Comparison of Efficacy with MAO-B Inhibitors

1. **Symptomatic Relief:**

While alternative therapies can provide symptomatic relief, MAO-B inhibitors like Selegiline and Rasagiline are often favored for their direct impact on dopamine metabolism and their neuroprotective effects. Understanding the strengths and limitations of each treatment option is essential for developing a comprehensive management plan.

2. **Individualized Treatment Plans:**

A multi-faceted approach that includes pharmacological alternatives, natural supplements, and lifestyle interventions can provide patients with personalized care tailored to their specific symptoms, preferences, and overall health status.

Conclusion

The landscape of dopaminergic therapies extends beyond MAO-B inhibitors, with various pharmacological alternatives and natural supplements available to support dopamine levels and improve patient outcomes. Additionally, non-pharmacological interventions can enhance overall health and well-being.

In the next chapter, we will explore strategies for optimizing dopamine health, focusing on approaches that individuals can adopt to maximize the benefits of their treatment regimen and enhance their quality of life. This holistic perspective will underscore the importance of a comprehensive approach to managing dopamine-related disorders.

Chapter 18: Strategies for Dopamine Optimization

Introduction

Optimizing dopamine levels is crucial for maintaining mental health, motivation, cognitive function, and overall well-being. This chapter outlines practical strategies for maximizing dopaminergic health, integrating pharmacological treatments with lifestyle modifications, dietary choices, and holistic practices. By adopting a comprehensive approach, individuals can enhance the effectiveness of MAO-B inhibitors like Selegiline and Rasagiline while promoting their overall health.

Pharmacological Strategies

1. **Adherence to Prescribed Treatments:**

   Consistently taking prescribed medications is fundamental to optimizing dopamine levels. Patients should work closely with their healthcare providers to ensure that they understand their treatment plans and the importance of adherence to maximize therapeutic benefits.

2. **Regular Monitoring and Follow-ups:**

   Regular check-ups with healthcare providers are essential for assessing treatment efficacy and making necessary adjustments. Monitoring for potential side effects and interactions can help maintain optimal dosing and enhance treatment outcomes.

3. **Combination Therapy:**

Discuss with healthcare providers the possibility of combining MAO-B inhibitors with other dopaminergic agents, such as dopamine agonists or COMT inhibitors. Such combinations can enhance overall dopaminergic activity and provide more comprehensive symptom relief.

Dietary Strategies

1. **Consume Dopamine-Boosting Foods:**

   ○ Incorporate foods rich in tyrosine and phenylalanine, such as lean meats, fish, eggs, dairy products, nuts, and legumes. These amino acids are crucial for dopamine synthesis.

   ○ Include antioxidant-rich foods like berries, dark chocolate, and green leafy vegetables to combat oxidative stress, supporting dopaminergic neuron health.

2. **Omega-3 Fatty Acids:**

   Include sources of omega-3 fatty acids, such as fatty fish, walnuts, and flaxseeds, in the diet. These fats support neuronal function and have been associated with improved mood and cognitive performance.

3. **Balanced Diet:**

   Emphasize a balanced diet with adequate vitamins and minerals, particularly B vitamins (B6, B12, and folate) and magnesium, which play a role in neurotransmitter synthesis and function.

4. **Hydration:**

Maintain proper hydration, as dehydration can negatively affect cognitive function and mood. Encourage a consistent intake of fluids throughout the day.

Exercise and Physical Activity

1. **Regular Physical Activity:**

   Engage in regular aerobic exercise, which has been shown to boost dopamine levels and improve mood. Aim for at least 150 minutes of moderate-intensity exercise each week, such as brisk walking, cycling, or swimming.

2. **Strength Training:**

   Incorporate strength training exercises at least two days a week. Resistance training can enhance dopamine receptor sensitivity and improve overall physical health.

3. **Incorporate Movement into Daily Routine:**

   Encourage the inclusion of physical activity in daily life, such as taking the stairs, walking during breaks, or participating in recreational sports. Small changes can contribute to overall dopamine enhancement.

Stress Management Techniques

1. **Mindfulness and Meditation:**

   Practice mindfulness meditation and relaxation techniques to reduce stress and promote emotional well-being. Mindfulness can help regulate mood and enhance dopamine function.

2. **Cognitive Behavioral Therapy (CBT):**

   Consider therapeutic approaches like CBT to address negative thought patterns and enhance coping strategies. Improving mental resilience can positively influence dopamine levels.

3. **Social Connections:**

   Foster positive social relationships and engage in social activities. Strong social support networks can enhance mood and contribute to better mental health, indirectly supporting dopaminergic function.

Holistic Practices

1. **Integrative Approaches:**

   Explore complementary therapies such as yoga, tai chi, or acupuncture. These practices may reduce stress and promote relaxation, positively influencing dopamine levels.

2. **Quality Sleep:**

   Prioritize good sleep hygiene. Aim for 7-9 hours of quality sleep each night, as sleep deprivation can adversely affect dopamine receptor sensitivity and overall cognitive function.

3. **Limit Substance Use:**

   Reduce or eliminate the use of substances that can negatively impact dopamine levels, such as excessive alcohol consumption and recreational drugs. Encouraging a healthy lifestyle is vital for optimizing overall health.

Conclusion

Optimizing dopamine levels involves a multifaceted approach that includes pharmacological treatments, dietary choices, physical activity, stress management, and holistic practices. By adopting these strategies, individuals can enhance the effectiveness of MAO-B inhibitors like Selegiline and Rasagiline and improve their overall quality of life.

In the next chapter, we will delve into case studies and patient outcomes, illustrating real-world applications and success stories related to the use of MAO-B inhibitors and lifestyle modifications. This exploration will provide valuable insights into the practical benefits of a comprehensive approach to dopaminergic health.

Chapter 19: Case Studies and Patient Outcomes

Introduction

Real-world applications of MAO-B inhibitors, such as Selegiline and Rasagiline, provide valuable insights into their effectiveness and impact on patients' lives. This chapter presents case studies illustrating the varied experiences of individuals using these medications, highlighting both successes and challenges. By examining these real-world outcomes, we can better understand how MAO-B inhibitors function in different contexts and the potential for optimizing their use in clinical practice.

Case Study 1: Selegiline in Early Parkinson's Disease

Patient Background:

- **Name**: John
- **Age**: 62
- **Diagnosis**: Early-stage Parkinson's disease (diagnosed 1 year prior)

Treatment Plan:

John was prescribed Selegiline as a first-line treatment to manage his mild motor symptoms, including slight tremors and rigidity. He was also advised on lifestyle modifications, including exercise and a balanced diet.

Outcome:

Over six months, John experienced significant improvement in his motor symptoms. His tremors decreased, and he reported better mobility and quality of life. Additionally, he adhered to his exercise regimen, which contributed to his overall well-being. Follow-up assessments indicated that he maintained stable dopamine levels and had no significant side effects from Selegiline.

Reflection:

John's case demonstrates the effectiveness of Selegiline in early Parkinson's management. His positive response underscores the importance of integrating pharmacological treatment with lifestyle interventions to enhance overall therapeutic outcomes.

Case Study 2: Rasagiline and Mood Disorders

Patient Background:

- **Name**: Maria
- **Age**: 58
- **Diagnosis**: Parkinson's disease with concurrent major depressive disorder

Treatment Plan:

Maria was prescribed Rasagiline to manage her Parkinson's symptoms and was also started on an SSRI for depression. Her treatment plan included regular psychotherapy sessions to support her mental health.

Outcome:

After three months, Maria reported improvements in both her motor symptoms and mood. Rasagiline helped manage her rigidity and bradykinesia while the SSRI alleviated her depressive symptoms. Her healthcare team noted that the combination therapy appeared to enhance her quality of life significantly.

Reflection:

This case highlights the benefits of using Rasagiline not only for motor symptoms in Parkinson's disease but also for its potential impact on mood. The collaborative approach of combining medications and therapy proved effective for Maria.

Case Study 3: Challenges with Tolerance

Patient Background:

- **Name**: David
- **Age**: 71
- **Diagnosis**: Advanced Parkinson's disease

Treatment Plan:

David had been on Selegiline for over three years. As his disease progressed, he experienced motor fluctuations and increased difficulty managing "off" episodes.

Outcome:

Despite initially positive responses to Selegiline, David developed tolerance, leading to diminished efficacy. His neurologist decided to adjust his treatment plan by increasing the dosage of Selegiline and adding a dopamine agonist to his regimen.

Reflection:

David's case illustrates the potential for tolerance development with long-term MAO-B inhibitor use. It emphasizes the need for ongoing assessment and treatment adjustment in patients with progressive conditions to maintain therapeutic efficacy.

Case Study 4: Selegiline and Cognitive Benefits

Patient Background:

- **Name**: Susan
- **Age**: 65
- **Diagnosis**: Early Alzheimer's disease with mild cognitive impairment

Treatment Plan:

Susan was prescribed Selegiline as part of her treatment strategy to address cognitive decline and improve mood. Her regimen also included dietary adjustments and cognitive training exercises.

Outcome:

After six months, Susan showed notable improvements in cognitive function, particularly in memory recall and attention. Her family reported positive changes in her mood and engagement in social activities, contributing to a more active lifestyle.

Reflection:

Susan's experience suggests that Selegiline may offer cognitive benefits beyond its dopaminergic effects, particularly in patients with early cognitive decline. This case underlines the importance of a holistic approach to treatment.

Conclusion

These case studies underscore the versatility and impact of MAO-B inhibitors in managing not only Parkinson's disease but also other neurological conditions and mood disorders. Each patient's journey reflects the importance of individualized treatment plans that combine pharmacological interventions with lifestyle adjustments and supportive therapies.

In the next chapter, we will explore emerging MAO-B inhibitors, examining their development, potential breakthroughs, and implications for future therapeutic options in dopamine-related disorders. This exploration will provide insights into the evolving landscape of dopaminergic therapies and their significance in clinical practice.

Chapter 20: Emerging MAO-B Inhibitors

Introduction

As research in neuropharmacology continues to evolve, new MAO-B inhibitors are being developed to enhance the treatment landscape for dopamine-related disorders. These emerging therapies aim to provide improved efficacy, safety profiles, and broader applications in clinical practice. This chapter reviews the latest developments in MAO-B inhibitors, potential breakthroughs, and their implications for future therapeutic options.

Overview of Newer MAO–B Inhibitors

1. **Novel Compounds in Development:**

A variety of novel compounds targeting the MAO-B enzyme are currently under investigation. These new agents aim to offer enhanced selectivity, longer duration of action, and improved safety profiles compared to existing MAO-B inhibitors like Selegiline and Rasagiline.

2. **Mechanism of Action:**

Emerging MAO-B inhibitors may employ unique mechanisms to achieve their therapeutic effects. Some compounds are designed to provide both MAO-B inhibition and neuroprotective effects, potentially reducing oxidative stress and inflammation in dopaminergic neurons.

Potential Breakthroughs in MAO-B Inhibition

1. **Dual Action Therapies:**

   Researchers are exploring dual-action therapies that not only inhibit MAO-B but also modulate other neurotransmitter systems, such as serotonin and norepinephrine. These therapies may offer a more comprehensive approach to treating conditions like depression and Parkinson's disease.

2. **Targeted Delivery Systems:**

   Advances in drug delivery systems, including nanotechnology, may enable targeted delivery of MAO-B inhibitors directly to the brain. This approach could enhance drug efficacy while minimizing systemic side effects, leading to improved patient outcomes.

3. **Long-Acting Formulations:**

The development of long-acting formulations of MAO-B inhibitors is underway. These formulations aim to provide sustained therapeutic effects with fewer doses, improving adherence and patient quality of life.

Clinical Trials and Research Directions

1. **Ongoing Clinical Trials:**

   Several clinical trials are underway to evaluate the safety and efficacy of new MAO-B inhibitors. These studies are crucial for determining the therapeutic potential of these agents in various conditions, including Parkinson's disease, Alzheimer's disease, and mood disorders.

2. **Exploring Efficacy Beyond Parkinson's Disease:**

   Research is increasingly focusing on the potential of MAO-B inhibitors in other neurodegenerative conditions and psychiatric disorders. Studies assessing their impact on cognitive function, mood stabilization, and overall quality of life are particularly promising.

3. **Combination Therapy Studies:**

Trials examining the use of emerging MAO-B inhibitors in combination with existing treatments (such as dopamine agonists or SSRIs) are also being conducted. This research aims to identify synergistic effects that could enhance overall treatment outcomes.

Implications for Clinical Practice

1. **Broader Treatment Options:**

   The emergence of new MAO-B inhibitors expands the treatment options available for patients with dopamine-related disorders. Clinicians will have access to a wider array of pharmacological agents, enabling more personalized treatment approaches.

2. **Enhanced Patient Outcomes:**

   Improved efficacy and safety profiles of new MAO-B inhibitors may lead to better patient adherence and satisfaction. The ability to tailor treatment regimens based on individual patient needs can enhance overall therapeutic success.

3. **Integrating Emerging Therapies:**

As new therapies become available, healthcare providers must stay informed about the latest developments in MAO-B inhibition. This knowledge will be essential for integrating these emerging therapies into clinical practice effectively.

Conclusion

Emerging MAO-B inhibitors represent a promising advancement in the treatment of dopamine-related disorders. With ongoing research and development, these therapies have the potential to enhance the management of conditions like Parkinson's disease, Alzheimer's disease, and mood disorders. As we continue to explore the therapeutic landscape, the integration of novel MAO-B inhibitors will be crucial for optimizing patient outcomes and advancing our understanding of dopaminergic health.

In the next chapter, we will explore the cognitive enhancement potential of MAO-B inhibitors, discussing the evidence from clinical trials and the implications for their use as nootropic agents. This exploration will provide insights into how these medications can contribute to cognitive health and overall well-being.

Chapter 21: MAO–B Inhibitors and Cognitive Enhancement

Introduction

Cognitive health is increasingly recognized as a critical component of overall well-being, particularly in the context of neurodegenerative diseases and mental health disorders. MAO-B inhibitors, such as Selegiline and Rasagiline, have been studied not only for their role in managing motor symptoms of Parkinson's disease but also for their potential cognitive-enhancing effects. This chapter explores the evidence supporting the cognitive benefits of MAO-B inhibitors, mechanisms underlying these effects, and their implications for therapeutic strategies.

Cognitive Benefits Observed in Clinical Trials

1. **Cognitive Function in Parkinson's Disease:**

   Several clinical trials have investigated the effects of MAO-B inhibitors on cognitive function in patients with Parkinson's disease. Results suggest that both Selegiline and Rasagiline may have neuroprotective effects that extend beyond motor control, potentially delaying cognitive decline.

2. **Meta-Analyses and Systematic Reviews:**

   Meta-analyses have indicated that MAO-B inhibitors can lead to improvements in various cognitive domains, including attention, executive function, and memory. These improvements may be particularly beneficial in the early stages of Parkinson's disease, where cognitive symptoms often begin to emerge.

3. **Assessment Tools:**

Cognitive assessments in studies typically include standardized scales such as the Mini-Mental State Examination (MMSE) and the Parkinson's Disease Cognitive Rating Scale (PDCRS). Findings consistently show that patients receiving MAO-B inhibitors exhibit better cognitive performance compared to those on placebo or alternative therapies.

Mechanisms of Cognitive Enhancement

1. **Dopaminergic Modulation:**

MAO-B inhibitors increase the availability of dopamine in the brain by preventing its breakdown. Enhanced dopaminergic signaling is crucial for cognitive functions, particularly in areas of the brain associated with learning, memory, and executive functions.

2. **Neuroprotection:**

By reducing oxidative stress and inflammation, MAO-B inhibitors may help protect neurons from damage. This neuroprotective action can preserve cognitive function and potentially slow the progression of cognitive decline associated with neurodegenerative diseases.

3. **Enhanced Neuroplasticity:**

   Research suggests that dopaminergic activity is linked to neuroplasticity—the brain's ability to adapt and reorganize itself. MAO-B inhibitors may facilitate neuroplastic changes that support learning and memory, contributing to improved cognitive performance.

4. **Impact on Other Neurotransmitter Systems:**

   Emerging evidence indicates that MAO-B inhibitors may also affect other neurotransmitter systems, including serotonin and norepinephrine, which are involved in mood regulation and cognitive processes. This multifaceted approach can enhance overall cognitive health.

Potential as a Nootropic

1. **Defining Nootropics:**

Nootropics are substances that may enhance cognitive function, particularly executive functions, memory, creativity, or motivation in healthy individuals. While traditionally used in the context of nootropic agents, there is growing interest in the use of MAO-B inhibitors for cognitive enhancement.

2. **Research in Healthy Populations:**

Preliminary studies have explored the cognitive effects of MAO-B inhibitors in healthy adults. Findings suggest that Selegiline and Rasagiline may improve aspects of cognitive performance, such as attention and working memory, although further research is necessary to establish their safety and efficacy in this population.

3. **Cognitive Resilience:**

The potential for MAO-B inhibitors to promote cognitive resilience in aging populations is an area of active research. Understanding how these medications may mitigate age-related cognitive decline could inform preventive strategies for maintaining cognitive health.

Clinical Implications and Recommendations

1. **Integrating Cognitive Health in Treatment Plans:**

Clinicians should consider the cognitive-enhancing potential of MAO-B inhibitors when developing treatment plans for patients with Parkinson's disease and other neurodegenerative disorders. This approach can lead to more comprehensive care that addresses both motor and cognitive symptoms.

2. **Patient Education:**

Educating patients and caregivers about the cognitive benefits of MAO-B inhibitors can help set realistic expectations regarding treatment outcomes. It can also encourage adherence to prescribed therapies.

3. **Monitoring Cognitive Function:**

Regular assessments of cognitive function should be included in the management of patients on MAO-B inhibitors. This ongoing evaluation can help detect changes in cognition and guide treatment adjustments as needed.

Conclusion

MAO-B inhibitors, particularly Selegiline and Rasagiline, show promise not only in managing motor symptoms but also in enhancing cognitive function in patients with Parkinson's disease and potentially other neurological conditions. The mechanisms underlying these cognitive benefits, including dopaminergic modulation, neuroprotection, and enhanced neuroplasticity, provide a compelling rationale for their use in clinical practice.

As research continues to unfold, the role of MAO-B inhibitors in cognitive enhancement may expand, offering new avenues for optimizing treatment in both clinical and healthy populations.

In the next chapter, we will outline practical guidelines for the use of MAO-B inhibitors, including dosage recommendations, titration strategies, and practical advice for clinicians and patients. This chapter aims to provide actionable insights to ensure the safe and effective use of these important medications in managing dopamine-related disorders.

Chapter 22: Practical Guidelines for Use

Introduction

As the understanding of MAO-B inhibitors, particularly Selegiline and Rasagiline, continues to evolve, it is essential to establish practical guidelines for their use in clinical practice. This chapter provides comprehensive recommendations on dosage, titration, patient education, and safety considerations to optimize treatment outcomes for individuals with dopamine-related disorders.

Dosage Recommendations

1. **Initial Dosing:**

   - **Selegiline:**

     The typical starting dose for Selegiline is

     5 mg once daily

     for oral tablets. For the transdermal patch
     formulation, the initial dose may vary based on the
     specific product but typically starts at

     6 mg/24 hours

     .

   - **Rasagiline:**

     Rasagiline is generally started at a dose of

     1 mg once daily

     . This dosing is effective in managing Parkinson's
     symptoms and is well-tolerated in most patients.

2. **Maintenance Dosing:**

○ **Selegiline:**

The dose may be increased to

10 mg daily

if needed, depending on patient response and tolerability. The transdermal patch may also be titrated up based on clinical response.

○ **Rasagiline:**

The maximum recommended dose for Rasagiline is

1 mg daily

, as higher doses do not significantly increase efficacy and may raise the risk of side effects.

3. **Special Populations:**

- **Elderly Patients**: Consider starting with lower doses and adjusting based on tolerance, as older adults may be more sensitive to medications.
- **Patients with Hepatic Impairment**: Caution should be exercised when prescribing MAO-B inhibitors to patients with liver dysfunction, as metabolism may be altered. Dose adjustments may be necessary.

Titration Strategies

1. **Gradual Adjustments:**

Titration should be done gradually, typically allowing

2-4 weeks
between dose adjustments to assess the therapeutic effect and monitor for side effects.

2. **Monitoring Response:**

Clinicians should evaluate patient response and tolerability at each stage of titration. This includes assessing both motor and non-motor symptoms in patients with Parkinson's disease.

3. **Maximal Dose Considerations:**

Avoid exceeding the maximum recommended doses for both Selegiline and Rasagiline unless under specialized circumstances, as higher doses do not guarantee improved efficacy and increase the risk of adverse effects.

Patient Education and Counseling

1. **Understanding the Medication:**

   Patients should be educated about the purpose of MAO-B inhibitors, how they work, and the expected benefits. This understanding can improve adherence to the treatment regimen.

2. **Recognizing Side Effects:**

   Inform patients about potential side effects, including nausea, dizziness, insomnia, and possible interactions with other medications, especially serotonergic agents. Education on what symptoms to report to their healthcare provider is crucial.

3. **Dietary Considerations:**

While MAO-B inhibitors are less restrictive than MAO-A inhibitors in terms of dietary restrictions, patients should still be advised to maintain a balanced diet. Specific emphasis should be placed on avoiding excessive intake of tyramine-rich foods, particularly for patients who may be sensitive.

4. **Monitoring for Interactions:**

Patients should be made aware of potential interactions with other medications, especially SSRIs, SNRIs, and other antidepressants. Encourage them to report all medications, including over-the-counter drugs and supplements.

Safety Considerations

1. **Adverse Effects Monitoring:**

   Regular follow-up appointments should be scheduled to monitor for side effects and assess therapeutic effectiveness. Blood pressure monitoring may be warranted, particularly in patients with a history of hypertension.

2. **Contraindications:**

   Contraindications for MAO-B inhibitors include severe hepatic impairment, concomitant use of certain medications (such as other MAO inhibitors or certain antidepressants), and known hypersensitivity to the components of the formulation.

3. **Emergency Situations:**

Educate patients about potential emergency situations, such as serotonin syndrome, which can occur with the combination of MAO-B inhibitors and serotonergic medications. Patients should seek immediate medical attention if they experience symptoms such as confusion, rapid heart rate, or severe changes in blood pressure.

Conclusion

Establishing practical guidelines for the use of MAO-B inhibitors is essential for optimizing treatment in patients with dopamine-related disorders. By adhering to recommended dosages, utilizing careful titration strategies, and prioritizing patient education, healthcare providers can enhance treatment outcomes and minimize risks.

In the next chapter, we will explore future directions in dopaminergic therapies, focusing on advances in neuropharmacology and potential innovations that could further impact the management of dopamine-related conditions. This exploration will shed light on the evolving landscape of treatment options and the promise they hold for enhancing patient care.

Chapter 23: Future Directions in Dopaminergic Therapies

Introduction

As our understanding of the dopaminergic system deepens, the landscape of dopamine-related therapies is continually evolving. This chapter explores the future directions of dopaminergic therapies, focusing on the advancements in neuropharmacology, innovative therapeutic strategies, and the potential for new treatments that target dopamine regulation. By examining emerging research and technologies, we can gain insights into the promising future of managing dopamine-related disorders.

Advances in Neuropharmacology

1. **Targeted Drug Delivery:**

   Innovative drug delivery systems, such as nanoparticles and liposomes, are being developed to enhance the targeting and efficacy of dopaminergic medications. These systems can facilitate crossing the blood-brain barrier more effectively, allowing for higher concentrations of drugs in the central nervous system while minimizing systemic side effects.

2. **Biologics and Gene Therapy:**

   The exploration of biologics, including monoclonal antibodies and gene therapy, presents new avenues for addressing dopamine-related disorders. Gene therapy approaches aim to modify or replace defective genes involved in dopamine synthesis or signaling, potentially providing long-term solutions for conditions like Parkinson's disease and certain forms of depression.

3. Combination Therapies:

Research is increasingly focusing on combination therapies that incorporate MAO-B inhibitors with other pharmacological agents, such as dopamine agonists, antioxidants, and neuroprotective agents. These multi-faceted approaches could enhance therapeutic outcomes by addressing various aspects of dopaminergic dysfunction.

Innovative Therapeutic Strategies

1. **Personalized Medicine:**

   The future of dopaminergic therapy lies in personalized medicine, where treatment is tailored based on an individual's genetic profile, disease stage, and specific symptomatology. Pharmacogenomics can help identify which patients are likely to benefit from particular medications, minimizing trial-and-error prescribing.

2. **Digital Health Solutions:**

   The integration of digital health technologies, such as mobile health apps and telemedicine, can facilitate ongoing monitoring of symptoms and treatment adherence. These tools allow for real-time feedback and adjustments to treatment plans, improving overall management of dopamine-related disorders.

3. Neurostimulation Techniques:

Emerging neurostimulation techniques, such as transcranial magnetic stimulation (TMS) and deep brain stimulation (DBS), are showing promise in modulating dopaminergic activity. These non-pharmacological approaches can provide symptomatic relief for patients who do not respond adequately to conventional medications.

New Treatment Modalities

1. **Novel MAO-B Inhibitors:**

   Ongoing research into new MAO-B inhibitors is likely to yield compounds with improved efficacy and safety profiles. These newer agents may offer unique mechanisms of action or longer-lasting effects, providing alternatives to existing therapies.

2. **Nootropic Agents:**

   The potential for nootropic agents to enhance cognitive function in both healthy and clinical populations is an area of growing interest. Research into compounds that affect dopamine pathways and cognitive enhancement will likely inform future therapeutic options.

3. **Lifestyle Interventions:**

As the role of lifestyle factors in dopaminergic health becomes clearer, future therapies will increasingly incorporate recommendations for diet, exercise, and stress management as complementary to pharmacological treatments. This holistic approach recognizes the interplay between physical health and dopaminergic function.

Challenges and Considerations

1. **Regulatory Hurdles:**

   New therapies, especially those involving innovative delivery systems or genetic modifications, face significant regulatory challenges. Navigating the complexities of drug approval processes will be crucial for bringing new treatments to market.

2. **Ethical Considerations:**

   The use of advanced therapies, particularly gene therapy and neurostimulation, raises ethical questions regarding long-term effects, informed consent, and equitable access to treatment. Addressing these concerns will be essential as these modalities develop.

3. **Patient-Centered Care:**

   Emphasizing patient-centered care remains critical as new therapies emerge. Engaging patients in shared decision-making about their treatment options fosters adherence and improves satisfaction with care.

Conclusion

The future of dopaminergic therapies is bright, with numerous advancements on the horizon. As research progresses, we can expect new treatments and strategies that enhance our ability to manage dopamine-related disorders effectively. By integrating innovative pharmacological approaches with lifestyle interventions and personalized medicine, healthcare providers can significantly improve the quality of life for individuals affected by these conditions.

In the concluding chapter, we will reflect on the key insights gained throughout this book and consider the evolving landscape of dopamine-focused treatments, emphasizing the importance of ongoing research and the commitment to advancing patient care.

Chapter 24: Conclusion: The Role of MAO-B Inhibitors in Dopamine Mastery

Introduction

As we conclude this comprehensive exploration of dopamine and the role of MAO-B inhibitors like Selegiline and Rasagiline, it is essential to reflect on the insights gained throughout this journey. Understanding the intricate relationships between dopamine synthesis, regulation, and the therapeutic potential of MAO-B inhibitors provides a solid foundation for future developments in the field of neuropharmacology. This chapter summarizes key findings, highlights the implications for clinical practice, and emphasizes the importance of continued research.

Key Insights

1. **Dopamine as a Central Player:**

   Dopamine is not just a neurotransmitter; it is a key player in mood regulation, motivation, and reward pathways. Its influence extends beyond motor control, affecting cognitive function and emotional well-being.

2. **MAO-B Inhibitors and Their Impact:**

   MAO-B inhibitors have emerged as significant therapeutic agents in managing Parkinson's disease, mood disorders, and cognitive decline. They enhance dopaminergic activity by preventing the breakdown of dopamine, thereby improving patient outcomes.

3. **Comprehensive Approach to Treatment:**

The effective use of MAO-B inhibitors necessitates a holistic approach that includes pharmacological intervention, lifestyle modifications, and patient education. Combining these strategies can optimize treatment outcomes and enhance the quality of life for individuals with dopamine-related disorders.

4. **Research and Innovation:**

Continuous advancements in neuropharmacology are paving the way for novel therapies and treatment strategies. The development of new MAO-B inhibitors, dual-action therapies, and personalized medicine approaches holds promise for improving the management of complex neurological conditions.

5. **Patient-Centered Care:**

Engaging patients in their treatment plans and educating them about their medications fosters adherence and enhances treatment efficacy. Recognizing the unique needs of each patient is critical in achieving optimal therapeutic outcomes.

Implications for Clinical Practice

1. **Integration of MAO-B Inhibitors:**

   Clinicians should consider incorporating MAO-B inhibitors into treatment regimens for patients with Parkinson's disease, Alzheimer's disease, and mood disorders. Understanding their pharmacological properties and potential benefits is crucial for effective management.

2. **Monitoring and Follow-Up:**

   Regular monitoring of patient progress and potential side effects is essential to ensure the safety and efficacy of treatment. Adjustments to dosages and treatment plans may be necessary based on individual responses.

3. **Emphasis on Lifestyle Factors:**

   Encouraging patients to adopt healthy lifestyle practices, including diet and exercise, can significantly impact their overall health and enhance the effects of MAO-B inhibitors.

Future Directions

1. **Continued Research:**

   Ongoing research into the mechanisms of dopamine regulation and the development of new MAO-B inhibitors is vital. Future studies should focus on the long-term effects of these therapies and their role in cognitive enhancement.

2. **Exploration of Combination Therapies:**

   Investigating the efficacy of combination therapies that include MAO-B inhibitors with other classes of medications could yield significant insights and improve treatment protocols.

3. **Public Awareness and Education:**

   Increasing public awareness about dopamine-related disorders and the benefits of MAO-B inhibitors can help reduce stigma and improve access to treatment. Education initiatives can empower patients and families to seek appropriate care.

Conclusion

MAO-B inhibitors have established themselves as crucial components in the management of dopamine-related disorders, offering therapeutic benefits that extend beyond symptom relief. As the field of neuropharmacology evolves, embracing innovative approaches and maintaining a focus on patient-centered care will be essential in optimizing treatment outcomes. The journey of mastering dopamine is ongoing, with the promise of new discoveries and advancements on the horizon.

In closing, the interplay between dopamine regulation, therapeutic interventions, and holistic health underscores the complexity of neurological health. By harnessing the potential of MAO-B inhibitors and committing to comprehensive care strategies, we can improve the lives of those affected by dopamine-related conditions, paving the way for a brighter future in neurohealth.

Chapter 25: Reflections on Mastering Dopamine: A Journey Forward

Introduction

The exploration of dopamine and the role of MAO-B inhibitors has illuminated the complexities of neurochemistry and its profound impact on mental health, motivation, and motor function. As we conclude "Mastering Dopamine; MAO-B Inhibitors Selegiline and Rasagiline," it is essential to reflect on the journey we've taken through the intricate mechanisms of dopamine regulation, the therapeutic potential of MAO-B inhibitors, and the implications for future treatment strategies. This chapter synthesizes key themes, lessons learned, and thoughts on how to navigate the evolving landscape of dopamine-focused treatments.

Key Themes and Insights

1. **The Centrality of Dopamine:**

Dopamine plays a crucial role as a neurotransmitter influencing various aspects of behavior, emotion, and cognition. Its significance in reward pathways, mood regulation, and motivation cannot be overstated. The interplay between dopamine levels and mental health highlights the need for effective management strategies for disorders related to dopaminergic dysfunction.

2. **Understanding MAO-B Inhibition:**

MAO-B inhibitors like Selegiline and Rasagiline represent a critical pharmacological approach to enhancing dopaminergic signaling. By preventing the breakdown of dopamine, these agents offer both symptomatic relief and potential neuroprotective benefits. Understanding their mechanisms has profound implications for the management of conditions such as Parkinson's disease and depression.

3. **Holistic Management Approaches:**

The treatment of dopamine-related disorders extends beyond pharmacotherapy. Integrating lifestyle factors, including diet, exercise, and mental well-being, can synergistically enhance therapeutic outcomes. Emphasizing a comprehensive approach to patient care fosters better adherence and improved quality of life.

4. **Patient-Centered Care:**

Engaging patients in their treatment journey and empowering them with knowledge about their condition and treatment options is essential. A patient-centered approach encourages collaboration between healthcare providers and patients, leading to more personalized and effective treatment plans.

5. **The Importance of Ongoing Research:**

The field of neuropharmacology is rapidly evolving, with ongoing research leading to new insights and innovative therapies. The exploration of novel MAO-B inhibitors, combination therapies, and neuroprotective strategies holds promise for enhancing treatment efficacy and expanding options for patients.

Future Considerations

1. **Integration of New Technologies:**

   The advancement of digital health technologies offers exciting possibilities for monitoring treatment outcomes and engaging patients in their care. Mobile applications, telemedicine, and wearable devices can facilitate real-time data collection, enhancing the management of dopamine-related disorders.

2. **Ethical Implications of New Therapies:**

   As new therapies emerge, including gene therapy and novel drug delivery systems, ethical considerations surrounding access, informed consent, and long-term effects must be addressed. Engaging with these ethical dimensions is vital for responsible advancements in treatment.

3. **Commitment to Lifelong Learning:**

The complexities of dopamine and its role in health necessitate a commitment to continuous education for healthcare providers. Staying informed about the latest research and treatment strategies is crucial for optimizing patient outcomes.

Conclusion: A Path Forward

The journey of mastering dopamine through the lens of MAO-B inhibitors has provided invaluable insights into the intricate relationship between neurochemistry and health. As we look to the future, the integration of scientific advancements, innovative treatment strategies, and a commitment to patient-centered care will be paramount in addressing the challenges posed by dopamine-related disorders.

In this evolving landscape, we must remain adaptable and open to new ideas, fostering a collaborative spirit in the pursuit of knowledge and better health outcomes. By embracing the complexities of dopamine and its regulatory mechanisms, we can enhance our understanding and approach to treatment, ultimately improving the lives of those affected by these conditions.

As we conclude this book, let us continue to explore, innovate, and inspire change in the realm of dopamine mastery. Together, we can unlock new possibilities for understanding and treating disorders of the dopaminergic system, paving the way for a brighter future in neurohealth.